PREDICTING O
OF INVESTMENTS IN
MAINTENANCE AND REPAIR
OF FEDERAL FACILITIES

M000251260

Committee on Predicting Outcomes of Investments in Maintenance and Repair for Federal Facilities

Board on Infrastructure and the Constructed Environment

Division on Engineering and Physical Sciences

NATIONAL RESEARCH COUNCIL
OF THE NATIONAL ACADEMIES

THE NATIONAL ACADEMIES PRESS
Washington, D.C.
www.nap.edu

THE NATIONAL ACADEMIES PRESS • 500 Fifth Street, N.W. • Washington, DC 20001

NOTICE: The project that is the subject of this report was approved by the Governing Board of the National Research Council, whose members are drawn from the councils of the National Academy of Sciences, the National Academy of Engineering, and the Institute of Medicine. The members of the committee responsible for the report were chosen for their special competences and with regard for appropriate balance.

This report was supported by a series of contracts between the National Academy of Sciences and the sponsor agencies of the Federal Facilities Council. Any opinions, findings, conclusions, or recommendations expressed in this publication are those of the authors and do not necessarily reflect the views of the organizations or agencies that provided support for the project.

International Standard Book Number-13:   978-0-309-22186-3
International Standard Book Number-10:   0-309-22186-2

Additional copies of this report are available from The National Academies Press, 500 Fifth Street, N.W., Lockbox 285, Washington, DC 20055; (800) 624-6242 or (202) 334-3313 (in the Washington metropolitan area); Internet, http://www.nap.edu.

Printed in the United States of America

THE NATIONAL ACADEMIES
Advisers to the Nation on Science, Engineering, and Medicine

The **National Academy of Sciences** is a private, nonprofit, self-perpetuating society of distinguished scholars engaged in scientific and engineering research, dedicated to the furtherance of science and technology and to their use for the general welfare. Upon the authority of the charter granted to it by the Congress in 1863, the Academy has a mandate that requires it to advise the federal government on scientific and technical matters. Dr. Ralph J. Cicerone is president of the National Academy of Sciences.

The **National Academy of Engineering** was established in 1964, under the charter of the National Academy of Sciences, as a parallel organization of outstanding engineers. It is autonomous in its administration and in the selection of its members, sharing with the National Academy of Sciences the responsibility for advising the federal government. The National Academy of Engineering also sponsors engineering programs aimed at meeting national needs, encourages education and research, and recognizes the superior achievements of engineers. Dr. Charles M. Vest is president of the National Academy of Engineering.

The **Institute of Medicine** was established in 1970 by the National Academy of Sciences to secure the services of eminent members of appropriate professions in the examination of policy matters pertaining to the health of the public. The Institute acts under the responsibility given to the National Academy of Sciences by its congressional charter to be an adviser to the federal government and, upon its own initiative, to identify issues of medical care, research, and education. Dr. Harvey V. Fineberg is president of the Institute of Medicine.

The **National Research Council** was organized by the National Academy of Sciences in 1916 to associate the broad community of science and technology with the Academy's purposes of furthering knowledge and advising the federal government. Functioning in accordance with general policies determined by the Academy, the Council has become the principal operating agency of both the National Academy of Sciences and the National Academy of Engineering in providing services to the government, the public, and the scientific and engineering communities. The Council is administered jointly by both Academies and the Institute of Medicine. Dr. Ralph J. Cicerone and Dr. Charles M. Vest are chair and vice chair, respectively, of the National Research Council.

www.national-academies.org

Preface

The federal government operates a portfolio of about 429,000 buildings and 482,000 other structures whose core purposes are to support the conduct of public policy, to help to defend the national interest, and to provide services to the U.S. public. Since 1990, studies have been issued, research has been undertaken, and technology has advanced in support of more strategic and more cost-effective management of federal facilities.

However, although progress has been made, major issues persist in regard to the maintenance and repair of federal facilities:

- Federal facilities continue to deteriorate.
- Federal agencies continue to operate and maintain facilities that are excess to their missions.
- Each federal agency approaches reinvestment in maintenance and repair differently.
- Federal facilities program managers have been unable to communicate effectively the link between reinvestment in facilities' maintenance and repair and the core missions of their agencies.
- The federal government as a whole has not taken a leadership role in the maintenance and repair of its facilities.

So, what is different now that merits a new look at and a new study about the maintenance and repair of federal facilities? In fact, much has changed in the last 10 years. Recognition of the importance of buildings that protect their occupants in the event of disaster arose in the wake of the 9/11 terrorist attacks. In the aftermath, new security standards, risk assessment and risk mitigation processes, and new technologies have been developed. Public debate about the reduction of greenhouse gas emissions associated with global climate change has brought the significance of buildings and their operations to the forefront, because the electricity used by buildings accounts for 40 percent of U.S. greenhouse gas emissions. Facilities also use substantial amounts of the nation's energy, water, and materials, and contribute to air and water pollution. Accordingly, they are an important factor in achieving—or not achieving—public policy goals for energy security and environmental sustainability. Because most of the buildings and other facilities used today will still be in use 30 years from now, better processes for operating and maintaining facilities will be essential if we are to achieve substantial reductions in greenhouse gas emissions, energy use, and water use.

In 2011, the federal operating environment of increasing fiscal constraints and a growing national debt provides an additional impetus to reexamine how investments in federal facilities are made. Operating and maintaining unneeded facilities constitute a drain on the federal budget and result in lost opportunities to strategically invest to improve the condition of those facilities that support current missions, to reduce energy and water use, and to meet other public policy objectives. Strategic investments in maintenance and repair activities can also result in economic benefits when products, supplies, and equipment are purchased and when federal agencies contract out maintenance and repair activities to private-sector firms.

For those reasons and others, the Federal Facilities Council asked the National Research Council to appoint an ad hoc committee of experts to develop methods, strategies, and procedures to predict outcomes of investments in maintenance and repair of federal facilities. The committee appointed to undertake that task, the Committee on Predicting Outcomes of Investments in Maintenance and Repair for Federal Facilities, was composed of experts from public, private, and academic organizations who had a wealth of experience in addressing the complex and diverse issues surrounding facilities management. The committee reviewed previous reports that focused on federal facilities management; held discussions with representatives of private-sector organizations, professional societies, and numerous federal agencies; and conducted research on specific relevant topics to formulate its findings and recommendations.

Based on its work, the committee concluded that new, more proactive, and more transparent approaches to the maintenance and repair of federal facilities are needed. New approaches will have to identify specific outcomes that can result from a given level of maintenance and repair investment and identify the risks—the probability of adverse consequences—associated with a lack of investment. Those approaches will help federal facilities managers and decision-makers to improve their targeting of investments to achieve multiple objectives and help them to manage risk.

Implementation of a more strategic, risk-based approach to investment in federal facilities maintenance and repair will require a continuous-improvement mind-set at all levels of government. Improvement goals and objectives should focus on the following:

- Ensuring that federal facilities are safe, secure, and in compliance with a host of health, safety, and environmental regulations,
- Disposing of excess facilities that no longer support agencies' missions and reducing the total federal facilities "footprint," and
- Operating mission-supportive facilities efficiently and effectively to reduce their overall costs and to support energy efficiency and other public policy objectives.

As a nation, we cannot continue to ignore the risks and potential consequences of under-maintaining federal facilities. During a period of decreasing budgets, downsizing, and increased competition for federal funding, the federal government and its agencies have the opportunity—and the responsibility—to implement new approaches to strategically reinvest in the portfolio of federal facilities. By taking a leadership role, the government can both address the deterioration of the nation's public assets and also help to achieve other public policy goals, such as energy security and sustainability.

David A. Skiven, *Chair*
Committee on Predicting Outcomes of
Investments in Maintenance and Repair for
Federal Facilities

Acknowledgments

The committee acknowledges the substantial contributions of members of the Federal Facilities Council and of Valerie Short of Jacobs Engineering, Michael Telson of the University of California, and Robert Moore of the National Institute of Standards and Technology (retired). The committee also thanks the following, who willingly and enthusiastically volunteered their time and ideas: Al Antelman, Valerie Baldwin, Steven Beattie, William Broglie, Karl Calvo, Douglas Christensen, James J. Dempsey, Andrew Dichter, Terrell Dorn, Maria Edelstein, Douglas Ellsworth, Michael Grussing, Dino Herrera, Jay Janke, Gerald Kokos, Peter Lufkin, Lance Marrano, Peter Marshall, William McNab, Lander Medlin, Joseph Morganti, Patrick Okamura, Peter O'Konski, Carl Rabenaldt, Dominic Savini, Robert St. Thomas, Kim Toufectis, Cynthia Vallina, Alex Willman, Stephen Wooldridge, Richard N. Wright, and John Yates.

This report has been reviewed in draft form by individuals chosen for their diverse perspectives and technical expertise, in accordance with procedures approved by the National Research Council's Report Review Committee. The purpose of this independent review is to provide candid and critical comments that will assist the institution in making its published report as sound as possible and to ensure that the report meets institutional standards for objectivity, evidence, and responsiveness to the study charge. The review comments and draft manuscript remain confidential to protect the integrity of the deliberative process. We thank the following individuals for their review of the report:

William W. Badger, Arizona State University,
William W. Brubaker, Hill International Inc. (retired),
Donald Coffelt, Carnegie Mellon University,

David G. Cotts, Consultant (retired),
Dennis P. Ferrigno, CAF & Associates, LLC,
Chris Hendrickson, Carnegie Mellon University,
Robert N. Jortberg, U.S. Navy (retired),
Richard G. Little, Keston Institute for Public Finance and Infrastructure
    Policy,
Judith Passwaters, E.I. du Pont de Nemours and Company,
Kumares C. Sinha, Purdue University,
Michael Vorster, Virginia Tech, and
Alan Washburn, U.S. Naval Postgraduate School.

Although the reviewers listed above have provided many constructive comments and suggestions, they were not asked to endorse the conclusions or recommendations, nor did they see the final draft of the report before its release. The review of this report was overseen by Lloyd Duscha, Department of Defense (retired). Appointed by the National Research Council, he was responsible for making certain that an independent examination of this report was carried out in accordance with institutional procedures and that all review comments were carefully considered. Responsibility for the final content of this report rests entirely with the authoring committee and the institution.

Contents

The Committee on Predicting Outcomes of Investments in Maintenance and Repair for Federal Facilities dedicates this report to our chair and valued colleague, David A. Skiven, who provided unwavering leadership and inspiration through all phases of the study process. He was a leader, gifted engineer, manager, mentor, and a tireless volunteer on behalf of the National Research Council, its Board on Infrastructure and the Constructed Environment, and the Engineering Society of Detroit Institute. Dave died shortly after the report was released to the public.

Summary

The deteriorating condition of federal facilities poses economic, safety, operational, and environmental risks to the federal government, to the achievement of the missions of federal agencies, and to the achievement of public policy goals. Primary factors underlying the deterioration are the age of federal facilities—about half are at least 50 years old—and decades of inadequate investment in their maintenance and repair. Those issues are not new and there are no quick fixes. However, the current operating environment provides both the impetus and the opportunity to place investments in maintenance and repair of federal facilities on a new, more sustainable course for the 21st century.

In 1990, the National Research Council Committee on Advanced Maintenance Concepts for Buildings found that "credible analyses indicate that we are systematically neglecting the maintenance of public facilities at all levels of government. We are spending our assets and wasting our inheritance" (NRC, 1990, p. ix). Thirteen years later, the U.S. Government Accountability Office (GAO) designated federal facilities as a "high-risk"[1] area because of long-standing problems with excess and underutilized facilities, deteriorating facilities, unreliable data, expensive space, and the threat of terrorism (GAO, 2003).

The problems persist in 2011. The federal government owns and leases about 429,000 buildings and an additional 482,000 structures (such as utility systems, roads and bridges, and miscellaneous military facilities) worldwide (GSA, 2010); they are valued at $1.26 trillion (GSA, 2006) to $1.5 trillion (GAO, 2008) and have annual operating costs of more than $47 billion (GAO, 2008). About $1.66 billion

---

[1]GAO's high-risk update is provided at the start of each new Congress. The high-risk reports are intended to help the new Congress "focus its attention on the most important issues and challenges facing the federal government" (GAO, 2003, p. 1).

of the annual operating costs is expended on 45,000 facilities that are reported to be excess or underutilized (GSA, 2010).

Despite the magnitude of that investment, funding for the maintenance and repair of federal facilities has been inadequate for many years, and myriad projects have been deferred. GAO has stated that the total backlog of deferred maintenance and repairs, which amounts to tens of billions of dollars, "may have a significant effect on future budget resources and our nation's long-term fiscal sustainability" (GAO, 2008, p. 4).

Continued underinvestment in maintenance and repair will lead to even greater deterioration and greater risk to the government.[2] Probable adverse events include system failures that will disrupt agencies' operations; higher operating and life cycle costs; hazards that lead to injuries and illnesses or loss of life and property; waste of water, energy, and other resources; operational inefficiencies; continued greenhouse gas emissions; greater fiscal exposure related to facilities ownership; and even greater backlogs of deferred maintenance and repairs.

Current and projected constraints on the federal budget and the rising national debt provide the impetus to reexamine all federal programs, activities, and operations to find more cost-effective ways to provide goods and services to the U.S. public. Several recent developments provide an opportunity and a foundation for implementing more strategic and more cost-effective investment practices for maintaining and repairing federal facilities.

One development is the recognition by both public-sector and private-sector organizations that well-managed facilities enable efficient operations and the achievement of organizational missions and objectives. Recognition of the multifaceted value of facilities has, in turn, resulted in more strategic facilities management practices that focus on entire portfolios of facilities and treat them as assets to organizations. Federal agencies have been implementing portfolio-based management processes, although the level of sophistication varies. With a few exceptions, agencies have not yet adopted more strategic, portfolio-based practices for linking maintenance and repair investments to their overarching missions.

A second development is the continued evolution of information and other technologies. Information tools and technologies are now available to monitor facilities' condition, energy use, and other performance dimensions and to collect data that can be used to measure and predict outcomes of maintenance and repair investments, to reduce long-term costs, to eliminate human error and bias, and to increase operational efficiencies. Information technologies also support telework, which is changing the concept of workplaces and the demand for physical space.

A third development is the federal government's recognition of the critical role of facilities in meeting the national challenges of energy independence, homeland security, environmental sustainability, and global climate change. In the

---

[2]The committee used Lowrance's definition of risk as "a measure of the probability and severity of adverse events" (Lowrance, 1976, p. 1).

United States today, facilities directly account for almost 40 percent of primary energy use, 12 percent of water use, and 60 percent of all nonindustrial waste (NSTC, 2008). The processes used to produce and deliver energy to facilities for heating, cooling, ventilation, computers, and appliances account for 40 percent of U.S. greenhouse gas emissions (NAS-NAE-NRC, 2008).

Congress and two presidential administrations have enacted legislation and issued other directives challenging federal agencies to take a leadership role in reducing their use of energy, water, and fossil fuels and in reducing their greenhouse gas emissions. Those goals will be met only through efficient and effective operation and maintenance of mission-supportive facilities, combined with an overall reduction in the amount of the total square footage ("footprint") of federal facilities. All those factors both require and enable changes in the approaches used to manage, maintain, and repair federal facilities.

Transforming the current portfolio of federal facilities into one that is more economically, physically, and environmentally sustainable at a time when budgets are decreasing is daunting. Nonetheless, this report identifies processes and practices for doing so.

STATEMENT OF TASK AND THE COMMITTEE'S APPROACH

In October 2009, the National Research Council appointed an ad hoc committee of experts to develop methods, strategies, and procedures to predict outcomes[3] anticipated from investments in federal facilities' maintenance and repair. The committee was asked to address the following questions:

- Are there ways to predict or quantify the outcomes that can be expected from a given level of investment in maintenance and repair of federal facilities or facilities' systems?
- What risks do deteriorating facilities, deteriorating building systems (such as mechanical and electrical) or deteriorating components (such as roofs and foundations) pose to the achievement of a federal agency's mission or to other organizational outcomes (for example, physical security, operating costs, worker recruitment and retention, and health care costs)?
- Do such risks vary by facility type (such as offices, hospitals, industrial, and laboratories), by system, or by function (such as research and administrative)? Can the risks be quantified?
- What strategies, measures, and data should be in place to determine the outcomes of facilities maintenance and repair investments? How can those strategies, measures, and data be used to improve the outcomes of investments?

---

[3]The committee used Webster's definition of an outcome as "something that follows as a result or consequence" (Webster's New Collegiate Dictionary, 1976).

- Are there effective communication strategies that federal facilities program managers can use to inform decision-makers better about the cost-effectiveness of levels of investment in facilities' maintenance and repair?

To fulfill its task, the committee (Appendix A) met five times from December 2009 to September, 2010, exchanged report chapters by e-mail, and held a series of conference calls. The committee reviewed previous NRC reports related to federal facilities management and gathered information from numerous federal agencies and several private-sector and professional organizations that were identified by the committee as industry leaders (Appendix B). The committee's findings and recommendations are based on the information gathered through the literature review, briefings, committee meetings, and the individual committee members' experience and expertise.

OVERALL CONCLUSIONS

Just as there are no quick fixes for issues related to the management of federal facilities, there is no simple answer to the question, What outcomes will result from a given level of investment in maintenance and repair of facilities? The answer will depend on a number of factors, including the specific mission and programs of an agency; the type, number, and distribution of the facilities used to enable missions and programs; the existing condition of those facilities; and the resources available for investment in maintenance and repair.

In the same vein, no single formula or equation is available that will quantify the relationships between a given level of investment and the types of outcomes that will result or the level of risk that will be mitigated. Instead federal facilities program managers, in concert with other federal facilities stakeholders, will need to work through a more complicated method that takes into account the many complexities of facilities management and investment. Nonetheless, it can be done. The result will be improved processes for and improved outcomes of investments in the maintenance and repair of federal facilities.

FINDINGS

Finding 1. An array of beneficial outcomes can be achieved through timely investments in facilities maintenance and repair (Table S.1). Those outcomes support mission achievement, compliance with regulations, improved condition, efficient operations, and stakeholder-driven initiatives. All the outcomes can be measured. Some outcomes including reliability and physical condition can be predicted; that is, they can be estimated before an investment is made or if an investment is not made.

Finding 2. Deteriorating facilities and systems pose risks to the federal government, its agencies, its workforce, and the public. Among them are risks to the

TABLE S.1 Beneficial Outcomes Related to Investments in Maintenance and Repair

Mission-Related Outcomes	Compliance-Related Outcomes	Condition-Related Outcomes	Efficient Operations	Stakeholder-Driven Outcomes
Improved reliability	Fewer accidents and injuries	Improved condition	Less reactive, unplanned maintenance and repair	Customer satisfaction
Improved productivity	Fewer building-related illnesses	Reduced backlog of deferred maintenance and repairs	Lower operating costs	Improved public image
Functionality	Fewer insurance claims, lawsuits, and regulatory violations		Lower life-cycle costs	
Efficient space utilization			Cost avoidance	
			Reduced energy use	
			Reduced water use	
			Reduced greenhouse gas emissions	

achievement of federal agencies' missions; risks to safe, healthy, and secure workplaces; risks to the government's fiscal soundness; risks to efficient and cost-effective operations; and risks to achieving public policy objectives.

Finding 3. The risks associated with deteriorating facilities vary by type of facility, by system, by existing condition, by function, by utilization, and, most important, by the relationship of facilities to an agency's mission. Risks can be identified qualitatively and some can be quantified.

Finding 4. Excess, underutilized, and obsolete facilities constitute a drain on the federal government's budget in costs and in forgone opportunities to invest in the maintenance and repair of mission-supportive facilities and to reduce energy use, water use, and greenhouse gas emissions.

Finding 5. To manage and mitigate the risks posed by the ownership of facilities, high-performance private-sector organizations do the following:

- Systematically dispose of excess and underutilized facilities.
- Pursue a proactive strategy to minimize their total facilities "footprint."
- Link maintenance and repair activities to the organization's business or mission and set priorities among them.

- Correlate the effects of systems-related failures with the business or mission.
- Correlate delays in timely maintenance and repair with sustainment cost.

Finding 6. To make the outcomes of and risks posed by investments in maintenance and repair projects and activities transparent to decision-makers at all levels of the organization, facilities managers in high-performance organizations do the following:

- Aggregate maintenance and repair requirements for some facilities' systems and components (such as life-safety systems and roofs) to provide for greater transparency and to identify operational efficiencies.
- Perform "knowledge-based" condition assessments; that is, tailor the frequency and level of inspection to the strategic importance of facilities and to the life cycle of systems and components to provide credible estimates of repair costs and remaining service lives.
- Measure outcomes as a basis of continuous improvement.
- Implement feedback systems to evaluate the performance of investments.

Finding 7. Investment strategies, definitions of maintenance and repair, maintenance and repair practices, and methods for budget development vary among federal agencies as a result of their different missions; the sizes, compositions, and distributions of their facilities; and their organizational cultures. The lack of common approaches makes it difficult to compare the effectiveness of maintenance and repair investments among federal agencies, to compare the benefits and pitfalls of different investment strategies, and to benchmark performance for the purpose of continuous improvement.

Finding 8. Reliable and appropriate data and information are essential for measuring and predicting outcomes of investments in federal facilities maintenance and repair. An array of data, tools, and technologies is available to support strategic decision-making, to quantify outcomes and risks by using empirical data, to expedite data collection, and to reduce human errors and bias.

Finding 9. Additional research and collaborative efforts are needed to continue to develop rapid and effective data-collection methods (such as the use of sensors and visual imaging devices), data definition and exchange standards that allow interoperability of data and software systems, and robust prediction models.

RECOMMENDATIONS

Recommendation 1 (Findings 4 and 5). To better manage the economic, physical, and environmental risks associated with facilities ownership, the federal government and its agencies should embark on a coordinated, funded, and sustained effort to dispose of excess and underutilized facilities. They should also proactively reduce their total facilities footprint through alternative work strategies and other measures.

Recommendation 2 (Findings 1, 5, and 6). Federal agencies should develop more strategic approaches for investing in facilities maintenance and repair to achieve beneficial outcomes and to mitigate risks. Such approaches should do the following:

- Identify and set priorities among the outcomes to be achieved through maintenance and repair investments and link them to achievement of agencies' missions and other public policy objectives.
- Provide a systematic approach to performance measurement, analysis, and feedback.
- Provide for greater transparency and credibility in budget development, decision-making, and budget execution.

Recommendation 3 (Findings 1, 2, and 3). To develop more strategic approaches to maintenance and repair investment, federal agencies should do the following:

- Identify and set priorities among the beneficial outcomes that are to be achieved through maintenance and repair investments, preferably in the form of a 5- to 10-year plan agreed on by all levels of the organizations. Elements of that type of plan are outlined in Chapter 7.
- Establish a risk-based process for setting priorities among annual maintenance and repair activities in the field and at the headquarters level. Guidance for doing that is contained in Chapter 7.
- Establish standard methods for gathering and updating data to provide credible, empirical information for decision support, to measure outcomes of investments in maintenance and repair, and to track and improve the results.

Recommendation 4 (Finding 6). Federal facilities program managers should plan for multiple internal and external communications when presenting maintenance and repair requests to other decision-makers and staff. The information communicated should be accurate, acknowledge uncertainties, and be available in multiple forms to meet the needs of different audiences. The basis of prediction of outcomes of a given level of investment in maintenance and repair should be transparent and available to decision-makers.

Recommendation 5 (Finding 7). Federal agencies and other appropriate organizations should continue to collaborate to develop and refine governmentwide measures for outcomes of maintenance and repair investments and to develop more standardized practices, unambiguous procedures, definitions, and models. The committee believes that those activities would be most effective if under the auspices of the Office of Management and Budget.

Recommendation 6 (Findings 6 and 8). Federal agencies should avoid the collection of data that serve no immediate mission-related purpose. Agencies should use a "knowledge-based" approach to condition assessment. Outcome metrics and models should make maximum use of existing data. When new or unique data are required to support the development of an outcome measure or model, there should be a clearly defined benefit to offset the cost of collecting and maintaining them.

Recommendation 7 (Findings 8 and 9). Federal agencies should continue to participate in and take advantage of collaborative efforts to develop rapid and effective data-collection methods (such as the use of sensors and visual imaging devices), to develop data-exchange standards that allow interoperability of data and software systems, to develop the empirical information needed for robust prediction models, and to develop practices that will reduce the cost of data collection and eliminate human error and bias.

1

Introduction

The U.S. federal government has a substantial and ongoing investment in its facilities, which include some 429,000 buildings and an additional 482,000 structures and infrastructure (GSA, 2010). The purpose of these facilities is to enable the achievement of federal agencies' missions, which include national defense; homeland security; international diplomacy; protecting the public's health, safety, and welfare; space exploration; fostering commerce; recreation; collecting and preserving historical and cultural artifacts and the arts; and scientific research. The facilities that enable those missions include military installations; embassy compounds; office and administrative space; satellite, communication, and data centers; hospitals; museums; laboratories; roads and bridges; dams and levees; inland waterways; power plants; and many other types of buildings, structures, and infrastructure.

The estimated replacement value of all federal government facilities (buildings, structures, and infrastructure) ranges from $1.26 trillion (GSA, 2006) to $1.7 trillion (GAO, 2008). Every year, the federal government spends as much as $47 billion to operate and maintain its facilities (GAO, 2008). Operating costs include energy costs, which fluctuate with the market but are always substantial. In fiscal year (FY) 2007, the latest year on which data are available, federal buildings used 392 trillion British thermal units (Btu) of energy at a cost of $6.5 billion (FEMP, 2010).

Despite the magnitude of the investment, many federal facilities are deteriorating because of decades of inadequate funding for their maintenance and repair, their age, and other factors. Deteriorating facilities, in turn, pose an array of risks to the achievement of federal agencies' missions, the achievement of public policy goals, and the federal government's fiscal soundness.

9

All indications are that funding for many federal agencies' programs will decrease in the near term because of current fiscal conditions and the rising national debt. This operating environment provides an impetus to reexamine the practices that have resulted in deteriorating, excess, and underutilized federal facilities and an opportunity to go forward in a more sustainable direction.

Because this report is intended for multiple audiences that have different backgrounds and interests, several key terms used in the report are explained here. If the committee has used a definition from another source, the source is cited (Box 1.1).

BOX 1.1
Terms Used in This Report

Customers as defined in this report are the users of federal facilities, including tenants and visitors.

Excess facilities are those which are no longer needed to support a federal agency's current or future missions.

Facilities refers to buildings (such as hospitals, barracks, embassies, and offices), other types of structures (such as parking, storage, and industrial), and infrastructure (such as power plants, water and sewer systems, railroads, roads, and bridges).

Federal facilities program managers are federal employees who are directly responsible for federal facilities programs; their responsibilities may include oversight of activities related to facilities design, construction, programming, budgeting, operations, maintenance, and evaluation.

Knowledge-based condition assessments use knowledge (quantifiable information) about a facility's systems and components to select the appropriate inspection type and schedule throughout its life cycle. Inspections are planned and executed on the basis of knowledge, not merely the calendar (Uzarski et al., 2007).

An *outcome* is something that follows as a result or consequence (Webster's New Collegiate Dictionary, 1976).

CHARACTERISTICS OF THE PORTFOLIO OF FEDERAL FACILITIES

The federal government owns about 429,000 buildings of many types (Figure 1.1) which have a total square footage (footprint) of 3.3 billion square feet (Table 1.1).

About 83 percent of the total square footage of federal buildings in the 50 states is owned space, 13 percent is leased, and 4 percent is managed otherwise (GSA, 2010).

In addition to buildings, the government owns 482,000 structures including utility systems, roads and bridges, parking, recreational and storage structures, and miscellaneous military facilities (Figure 1.2 and Table 1.2).

Portfolio-based facilities asset management is a systematic process of maintaining, upgrading, and operating physical assets cost effectively. It combines engineering principles with sound business practices and economic theory, and provides tools to facilitate a more organized, logical approach to decision-making. A facilities asset management approach allows for both program or network-level management and project-level management, and thereby supports both executive-level and field-level decision-making (NRC, 2004a, p. 32).

Public-private partnerships are contractual agreements between public and private-sector organizations in which the private sector, in exchange for compensation, agrees to deliver services, or even facilities, that could be provided by the public sector (Keston Institute, 2011).

Risk is a measure of the probability and severity of adverse events (Lowrance, 1976, p. 1).

Stakeholders in maintenance and repair of federal facilities include federal departments and agencies; facilities program managers; customers; oversight organizations such as the Office of Management and Budget; Congress; the administration; and the general public.

Total cost of ownership is the total of all expenditures an owner organization will make over the life cycle of a facility, that is, all expenditures related to planning, design, construction, operations and maintenance, renewal, revitalization, and disposal (NRC, 2008).

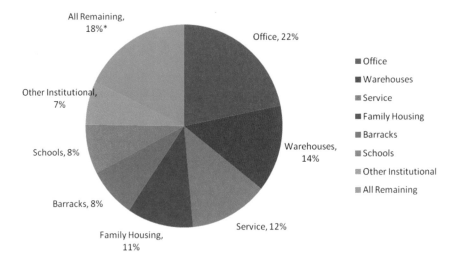

FIGURE 1.1 Federal buildings by predominant use in square feet as reported for FY 2009. NOTE: All remaining uses include prisons and detention centers, hospitals, laboratories, industrial, communication systems, museums, and post offices. SOURCE: GSA, 2010.

TABLE 1.1 Federal Buildings by Predominant Use and Square Footage as Reported for FY 2009

Predominant Use	Square Feet in Millions
Office	740.8
Warehouses	460.4
Services	416.2
Family housing	364.9
Barracks	271.2
Schools	251.7
Other institutional uses	221.4
All remaining uses	612.8
Total square feet	3,339.4

SOURCE: GSA, 2010.

Federal facilities (buildings and structures combined) are owned and managed by more than 30 departments and agencies. The Army, Navy, Air Force, Department of Veterans Affairs (VA), and the General Services Administration (GSA) manage the greatest numbers of buildings and structures and the greatest amounts of building space as measured by square footage (Table 1.3).

Much of the current federal facilities portfolio "reflects an infrastructure based on the business model and technological environment of the 1950s" and "many of

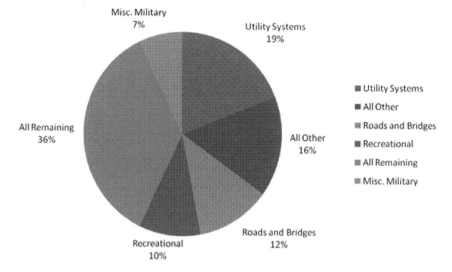

FIGURE 1.2 Predominant use of structures by number of assets as reported for FY 2008.
SOURCE: GSA, 2009.

TABLE 1.2 Predominant Use by Number of Structures as Reported in FY 2008

Predominant Use	Number of Structures
Utility systems	91,000
Roads and bridges	58,000
Recreational (other than buildings)	49,000
Parking structures	39,000
Miscellaneous military facilities	35,000
Storage (other than buildings)	30,000
Navigation and traffic aids (other than buildings)	26,000
Reclamation and irrigation	16,000
Communication systems	14,000
All other[a]	79,000
All remaining uses[b]	45,000
Total number of structures	482,000

[a]All other uses include those that are not captured in the predominant use categories.

[b]All remaining uses include airfield pavements, flood control and navigation, harbors and ports, industrial (other than buildings), monuments and memorials, museums, power development and distribution, railroads, research and development (other than laboratories), service (other than buildings), space exploration structures, and weapons ranges.
SOURCE: GSA, 2009.

TABLE 1.3 The Seven Agencies Managing the Greatest Amounts of Total Building Square Footage

Agency	Total Number of Buildings and Structures	Total Building Square Footage
U.S. Army	251,676	932,367,000
U.S. Air Force	134,788	606,191,000
U.S. Navy	150,576	578,305,000
GSA	9,213	407,941,000
VA	9,220	156,344,000
Department of Energy	18,354	129,239,000
Department of State	15,743	72,668,000
Total	589,570	2,883,055,000

SOURCE: GSA, 2010.

the assets are no longer effectively aligned with, or responsive to, agencies' changing missions. . . ." (GAO, 2003, p. 1). In FY 2008, federal agencies disposed[1] of about 25,000 facilities that were identified as excess with regard to current missions or as underutilized; they had annual operating costs of $119 million. In FY 2009, an additional 19,500 facilities with operating costs of $149 million were disposed of (GSA, 2010). Nonetheless, as of FY 2009, federal agencies reported that they still managed more than 45,000 facilities that are excess to their current missions or underutilized, and that they were spending more than $1.66 billion to operate them (GAO, 2011b).

The accumulation of excess and underutilized properties is a result of 200 years of acquiring facilities to support changing missions and new federal programs, and of the difficulty of disposing of facilities once they are acquired. Obstacles hindering sale, transfer of title, demolition or other methods of disposition include myriad regulations for transferring title to nonfederal entities, disincentives created by the federal budget structure, security issues related to the location of some excess facilities, and the condition of some facilities (NRC, 1998). Additional issues include the "numerous stakeholders that have an interest in how the federal government carries out its real property acquisition, management, and disposal practices" and a "complex legal environment that has a significant impact on real property decision making and may not lead to economically rational outcomes" (GAO, 2011b, p. 5).

LONG-STANDING INVESTMENT AND MANAGEMENT ISSUES

Excess and underutilized facilities are only one of several long-standing issues related to investment in and management of federal facilities. In 1990,

---

[1]Disposition methods include demolition, federal transfer, sale, public benefit conveyance, and others (GSA, 2010).

the National Research Council Committee on Advanced Maintenance Concepts for Buildings was asked to undertake a broad review of maintenance and repair activities of government agencies and to recommend how these activities might be improved (NRC, 1990). The committee found that, "credible analyses indicate that we are systematically neglecting the maintenance of public facilities at all levels of government. We are spending our assets and wasting our inheritance" (NRC, 1990, p. ix). One of its recommendations, which became a standard for public facilities management, was the following (NRC, 1990, p. xii):

> An appropriate budget allocation for routine M&R [maintenance and repair] for a substantial inventory of facilities will typically be in the range of 2 to 4 percent of the aggregate replacement value of those facilities (excluding land and major associated infrastructure). In the absence of specific information upon which to base the M&R budget, this funding level should be used as an absolute minimum value. Where neglect of maintenance has caused a backlog of needed repairs to accumulate, spending must exceed this minimum level until the backlog has been eliminated.

Some federal agencies have used, and continue to use, the 2 to 4 percent guideline for developing and justifying budget requests for maintenance and repair activities. However, no agency has reported actually investing in maintenance and repair activities at a level as high as 2 percent of the current replacement value of its portfolio of facilities and underinvestment remains an issue (FFC, 1996; committee briefings, 2009 and 2010).

In 2003, the U.S. Government Accountability Office (GAO) designated federal real property as a "high-risk"[2] topic because of long-standing problems with excess and underutilized facilities, deteriorating facilities, lack of reliable governmentwide data for strategic asset management, the high costs of leased space, and the costs and challenges of securing property against potential threat of terrorism. In its report GAO stated that "many assets are in an alarming state of deterioration; agencies have estimated restoration and repair needs to be in the tens of billions of dollars" (GAO, 2003, p. 1). Furthermore, current trends "have multibillion dollar cost implications and can seriously jeopardize mission accomplishment" (GAO, 2003, p. 1). For example, the Department of Defense (DOD) reported that many of its facilities were not adequate to meet the war-fighting and operational concepts of the 21st century and that commanders rated two-thirds of their infrastructure to be in such poor condition as to affect mission accomplishment and morale substantially (GAO, 2003). In a similar vein, GSA noted that some of its buildings had electrical systems that were not capable of handling 21st century technologies, "which is critical to tenant

---

[2]The GAO's high-risk update is provided at the start of each new Congress. The high-risk reports are intended to help the new Congress "focus its attention on the most important issues and challenges facing the federal government" (GAO, 2003, p. 1).

agencies' accomplishing their missions" (GAO, 2003, p. 23). The GAO (2003, p. 1) concluded that:

> Resolving these problems will require high-level attention and effective leadership by both Congress and the administration. Also, because of the breadth and complexity of the issues, the long-standing nature of the problems, and the intense debate that will likely ensue, current structures and processes may not be adequate to address the problems. Thus, there is a need for a comprehensive, integrated transformation strategy for real property.

In 2004, Executive Order 13327, *Federal Real Property Asset Management,*[3] provided direction for meeting some of those issues and established the Federal Real Property Council (FRPC). The FRPC is an interagency council composed of representatives of the 24 largest land-holding agencies and chaired by the Office of Management and Budget (OMB). From 2004 to the end of 2010, the FRPC issued guidance focused on improving the strategic management of federal buildings and structures, improving the management of the condition of facilities, developing asset management plans, implementing controls to improve the reliability of facilities-related data, and developing a set of government-wide performance measures related to the management of portfolios of facilities (GAO, 2011a).

Nonetheless, backlogs of deferred maintenance and repair projects continue to grow. The 2008 GAO report, *Federal Real Property: Government's Fiscal Exposure from Repair and Maintenance Backlogs is Unclear,* found important differences in how agencies plan, estimate, and fund maintenance and repair activities. At that time, GAO recommended that OMB, in conjunction with the FRPC and in consultation with the Federal Accounting Standards Advisory Board (FASAB)[4] explore the potential for developing a uniform reporting requirement that would capture the government's fiscal exposure from maintenance and repair backlogs, because "this exposure may have a significant effect on future budget resources and our nation's long-term fiscal sustainability" (GAO, 2008, p. 4). As of the date of release of the present report, a uniform reporting requirement has not yet been approved for use by federal agencies.

As of January 2011, federal real property remained on GAO's "high-risk" list because of continuing issues related to leasing practices, excess properties, and physical security (GAO, 2011b).

Additional long-standing issues that pose obstacles for effective investment in and management of federal facilities have been identified in two previous National Research Council studies. Among them are the following:

---

[3]The full text of the executive order is available at http://edocket.access.gpo.gov/2004/pdf/04-2773.pdf.

[4]The mission of the FASAB is to develop accounting standards after considering the financial and budgetary needs of congressional oversight groups, executive agencies, and the needs of other users of federal financial information.

- The focus on the first (design and construction) costs of facilities in the budget process as opposed to their life-cycle (long-term operations and maintenance) costs (NRC, 1998).
- Budgeting and accounting processes that create disincentives for cost-effective investments in maintenance and repair (NRC, 1998).
- The distributed nature of decision making about federal facilities investments and the short-term outlook of decision-makers, which result in a lack of accountability for stewardship (NRC, 1998; 2004a).

Two additional long-standing issues affect federal facilities management and investment. One is the variation in definitions related to maintenance and repair used by federal government agencies. For example, in briefings to the committee, the Department of Energy (DOE) and the National Oceanic and Atmospheric Administration (NOAA) reported that they define maintenance and repair together as

> Day to day work that is required to sustain property in a condition suitable for it to be used for its designated purposes, including preventive, predictive and corrective maintenance.

The Bureau of Overseas Buildings Operations of the Department of State defines maintenance and repair as

> Services and/or materials used for items of a recurring nature to prevent damage which would be more costly to restore than to prevent. Examples include painting, weather stripping and the preventive maintenance of building systems. A second category of M&R [maintenance and repair] services includes services and/or materials used for items of a minor nature such as repairs of broken pipes. A third category consists of bulk M&R supplies for use in Government-owned and long-term leased properties such as paint, lumber, plumbing supplies, and electrical wire.

And the U.S. Coast Guard defines maintenance, repair, and in-kind replacements (M), as

> Activities needed to keep a building or infrastructure operational, specifically focusing on physical continuity. These activities are required to achieve the full economic life of real property assets, components, assemblies, and systems. Also included are in-kind replacements of components and systems at the end of their economic life, such as a chiller for a chiller, or a roof for a roof. Energy retrofits motivated by economic considerations also fall into this category. Additional "M" work includes survey, inspection, and assessment work used to identify, scope, and schedule "M" activities.

Those definitions and the definitions used by other agencies are tailored to the methods of operating and the organizational culture of individual agencies, so, it is difficult to compare levels of maintenance and repair investment among federal agencies, to quantify those investments for the federal government as a whole, or to benchmark investments with non-federal organizations.

A second issue is the variation in budgeting, priority setting, and execution related to maintenance and repair investments. Agencies prepare their budget requests for maintenance and repair funding (and all other programs) 2 years in advance of the fiscal year in which the funding will be appropriated. Typically the identification of maintenance and repair projects that require funding begins at the field level. Methods for developing a total funding request vary. Some agencies take past budgets and increase them by some percentage to cover inflation, new program requirements, or a backlog of deferred maintenance projects. Others use a guideline such as the NRC's 2 to 4 percent of current replacement value, and still others a facilities sustainment model (FSM) developed for the DOD.

Project priority setting takes place at the organization level typically associated with approval authority, although ranking may also occur at different points as the lists of requirements work their way up the chain of command. Priorities for specific maintenance and repair projects that will cost less than some dollar amount (by statute or regulation) may be set at the local level, whereas priorities for more expensive projects may be set at the headquarters level. The process used to set priorities for projects may range from the discretion of an individual to a committee discussion and may encompass a subjective ad hoc approach, stakeholder requests, a very structured objective approach that uses matrices or algorithms to rank projects, or combinations thereof.

The lack of standardization or comparability in developing funding requests makes it difficult for individual agencies and the government as a whole to identify the beneficial outcomes or the adverse consequences of different investment strategies, to share lessons learned, and to improve the outcomes of maintenance and repair investments governmentwide.

IMPETUS FOR AND FOUNDATION OF
MORE SUSTAINABLE PRACTICES

Historically, obtaining funds to maintain the federal government's buildings and infrastructure has been a challenge (DOD, 2001; GAO, 2008). Senior executives and Congress are inundated with requests to support mandates and discretionary programs, each accompanied with compelling messages and evidence.

In 2011, the challenge to find support for federal facilities investment is probably greater than any time in the recent past because of the increasing national debt. The report *Choosing the Nation's Fiscal Future* (NAPA and NRC, 2010) shows that the federal government has been spending much more than it has been collecting in revenues and will do so for the foreseeable future if current policies are continued (Figure 1.3).

Because of the projected growth in federal spending for Medicare, Medicaid, and Social Security, the report concludes that any efforts to rein in future defi-

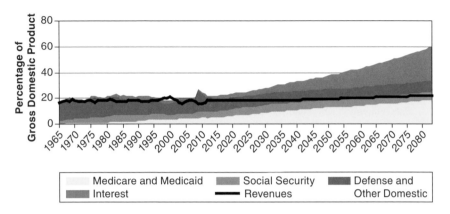

FIGURE 1.3 The long-term budget outlook. SOURCE: NAPA and NRC, 2010.

cits must entail either large increases in taxes to support these programs, major restraints on their growth, or some combination of the two (NAPA and NRC, 2010). It states that if the "choice is to keep the federal government's share of the economy close to the level of the past several decades, the government would have to scale back what it does, and extremely difficult choices would have to be made about what social goals to pursue less vigorously and what programs to end" (NAPA and NRC, 2010, p. 2). A January 2011 report by GAO reaches similar conclusions, noting that in the absence of policy changes, "the federal government faces a rapid and unsustainable growth in debt. . . . Addressing the long-term fiscal challenge . . . will likely require difficult decisions affecting both federal spending and revenue" (GAO, 2011c, p. 8).

The current fiscal situation provides the impetus for reexamining practices for federal facilities management, maintenance, and repair, and an opportunity to propose practices that will be more sustainable in the long term. The foundation of more sustainable practices is provided by portfolio-based facilities management, information tools and technologies, and recognition of the role of facilities in achieving public policy goals of energy independence and environmental sustainability.

Portfolio-Based Facilities Management

Recognition of the costs of facilities, of the role of facilities in enabling organization missions, of facilities' effects on occupants' health, safety, and security, and of facilities' effects on the environment, has been the impetus for more strategic management approaches by both private-sector and government organizations. One important change has been the shift from tactical concerns, tasks, and practices oriented to the operation of individual buildings and structures to a focus on the entire portfolio of facilities and integrated resource management (NRC, 2004a). Portfolio-based facilities management has been defined (NRC, 2004a, p. 32) as a

> systematic process of maintaining, upgrading, and operating physical assets cost effectively. It combines engineering principles with sound business practices and economic theory, and provides tools to facilitate a more organized, logical approach to decision making. A facilities asset management approach allows for both program or network-level management and project-level management and thereby supports both executive-level and field-level decision making.

A portfolio-based facilities management approach "allows organizations to integrate facilities considerations into corporate decision-making and strategic planning processes" (NRC, 2004a, p. 32). Thus, facilities are treated as valuable assets that contribute to the overall effectiveness of an organization.

Within the last 7 years, federal agencies have developed asset management plans. These plans are updated annually and are intended to "help agencies take a more strategic approach to real property [facilities and land] management by indicating how real property moves the agency's mission forward, outlining the agency's capital management plans, and describing how the agency plans to operate its facilities and dispose of unneeded real property, including listing current and future disposal plans" (GAO, 2011b, pp. 6, 7).

Federal agencies have been implementing portfolio-based management processes, but the level of sophistication varies. With a few exceptions, agencies have not yet adopted more strategic, portfolio-based practices for linking maintenance and repair investments to their organization's overarching mission.

Information Tools and Technologies

Many factors are driving a more strategic approach to facilities management and investment, and information tools and technologies are enabling it. Information tools and technologies are now available for monitoring facilities' condition, energy use, and other performance dimensions; for collecting data in "real time" to support strategic decision-making; for eliminating human error and bias; and for increasing operational efficiencies.

Because organizations can operate around the clock by having business units networked through the Internet and other technologies, the concept of workplace also is changing. Alternative work arrangements, such as telework enabled by technology, allow people to conduct work from home, airports, or other locations. The trend is changing the demand for centralized office space while also making the uninterrupted supply of power for telecommunications, cooling, and ventilation ever more critical. All those factors both enable and require changes in how federal facilities are managed, maintained, and repaired.

The Role of Facilities in Public Policy Issues

While federal agencies have been implementing new management practices, additional facilities-related goals have been established through legislation, executive orders, and presidential memorandums. The genesis of many of those goals is the ever-growing knowledge about the relationships between facilities and the natural environment; between indoor environments and the health, safety, and productivity of the people who use them; and about the magnitude of the costs and resources required to operate and maintain facilities. The amount of resources used by facilities and the costs of facilities have been noted previously. With respect to the health, safety, and productivity of building occupants, cause-effect relationships have been scientifically documented between waterborne pathogens in water systems and Legionnaire's disease and Pontiac fever; between microorganisms growing in contaminated ventilation and humidification systems and hypersensitivity pneumonitis and humidifier fever; between the release of carbon monoxide and carbon monoxide poisoning; between the presence of radon, secondhand smoke, and asbestos in buildings and lung cancer; and in connection with nonspecific symptoms—including eye, nose, and throat irritations—sometimes referred to as "sick-building syndrome" (FFC, 2005).

As the managers of the largest portfolio of facilities in the United States, federal agencies have been challenged to lead by example in operating their buildings and structures more sustainably over their life cycles. Laws have been enacted and other directives have been issued that establish goals of reducing the use of water, energy, and fossil fuels; improving indoor environmental quality; and reducing greenhouse gas emissions.

The Energy Independence and Security Act (EISA) of 2007, for example, defined the attributes of high-performance green buildings and established a set of goals and baselines for the reduction of energy, water, and fossil fuel use in federal buildings. The EISA standards address new construction, major renovations of existing structures, replacement of installed equipment, renovation, rehabilitation, expansion, and remodeling of existing space. More specifically, the EISA defined a high-performance green building as one that, during its life-cycle, as compared with similar buildings (as measured by Commercial Buildings Energy Consumption Survey data from the Energy Information Agency),

(A) Reduces energy, water, and material resource use.
(B) Improves indoor environmental quality, including reducing indoor pollution, improving thermal comfort, and improving lighting and acoustic environments that affect occupant health and productivity.
(C) Reduces negative impacts on the environment throughout the life-cycle of the building, including air and water pollution and waste generation.
(D) Increases the use of environmentally preferable products, including bio-based, recycled content, and nontoxic products with lower life-cycle impacts.
(E) Increases reuse and recycling opportunities.

(F) Integrates systems in the building.
(G) Reduces the environmental and energy impacts of transportation through building location and site design that support a full range of transportation choices for users of the building.
(H) Considers indoor and outdoor effects of the building on human health and the environment, including improvements in worker productivity, the life-cycle impacts of building materials and operations, and other factors considered to be appropriate.

Among other provisions, EISA requires that federal agencies reduce their total energy consumption by 30 percent by 2015, relative to 2003 consumption.

Executive Order 13423, *Strengthening Federal Environmental, Energy, and Transportation Management*, also issued in 2007, requires federal agencies to reduce their water intensity (gallons per square foot) by 2 percent each year through FY 2015 for a total of 16 percent relative to water consumption in FY 2007. It also requires federal agencies to ensure that 15 percent of the existing federal capital asset building inventory of each agency incorporate the sustainable practices outlined in the "Guiding Principles for Federal Leadership in High Performance and Sustainable Buildings" (hereinafter the Guiding Principles) by the end of FY 2015.[5]

The overall goals and objectives of the Guiding Principles are to reduce the total ownership cost of facilities; to improve energy efficiency and water conservation; to provide safe, healthy, and productive built environments; and to promote sustainable environmental stewardship. With respect to indoor environmental quality, the Guiding Principles recommend that agencies meet industry standards for ventilation, humidity, and temperature, and that they establish and implement a moisture control strategy to prevent building damage and mold contamination—actions that are related primarily to the efficient operation and maintenance of buildings.

Executive Order 13514, *Federal Leadership in Environmental, Energy, and Economic Performance,* issued in 2009, challenges federal agencies to lead by example in creating a clean energy economy and establishes more than 20 facilities-related goals for doing so. For the most part, agencies have not received additional funding to fulfill these mandates; instead, they must shift funding away from other programs and activities.

Because most facilities that federal agencies will be using for the next 20 to 30 years exist today, the primary methods for meeting those goals will be through efficient operations, maintenance, repair and retrofitting of existing facilities, and

---

[5]The Guiding Principles are the following: (1) Employ integrated design principles; (2) optimize energy performance; (3) protect and conserve water; (4) enhance indoor environmental quality; and (5) reduce environmental impact of materials. Available at http://www.energystar.gov/ia/business/Guiding_Principles.pdf.

the consolidation of the overall federal facilities footprint, not new construction (NRC, 2011).

Transforming the portfolio of federal facilities into one that is more economically, physically, and environmentally sustainable at the same time as budgets are being cut is daunting. To help provide direction in doing it, the Federal Facilities Council (FFC) [6] asked the National Research Council for its advice and assistance.

STATEMENT OF TASK

In October 2009, the National Research Council appointed an ad hoc committee of experts to develop methods, strategies, and procedures to predict outcomes anticipated from investments in federal facilities' maintenance and repair. The committee was also asked to address the following questions:

- Are there ways to predict or quantify the outcomes that can be expected from a given level of investment in maintenance and repair of federal facilities or facilities' systems?
- What risks do deteriorating facilities, deteriorating building systems (such as mechanical and electrical), or deteriorating components (such as roofs and foundations) pose to the achievement of a federal agency's mission or to other organizational outcomes (for example, physical security, operating costs, worker recruitment and retention, and health care costs)?
- Do such risks vary by facility type (such as offices, hospitals, industrial, and laboratories), by system, or by function (such as research and administrative)? Can the risks be quantified?
- What strategies, measures, and data should be in place to determine the outcomes of facilities maintenance and repair investments? How can those strategies, measures, and data be used to improve the outcomes of investments?
- Are there effective communication strategies that federal facilities program managers can use to inform decision-makers better about the cost-effectiveness of levels of investment in facilities' maintenance and repair?

THE COMMITTEE'S APPROACH

The committee members had expertise in facilities management, engineering, budgeting and finance, information technologies and data collection, the development of facilities-related models and performance measures, and risk identification, analysis, mitigation, and communication. The members have worked in

---

[6]The FFC is a cooperative association of more than 20 federal departments and agencies operating under the auspices of the National Research Council. The FFC's mission is to identify and advance technologies, processes, and management practices that improve the performance of federal facilities over their entire life cycle, from planning to disposal.

federal agencies, local governments, industry, and academia (see Appendix A for biosketches of committee members).

The committee began its work in December 2009 with a review of previous NRC reports on federal facilities management. Those reports included *Committing to the Cost of Ownership: Maintenance and Repair of Public Buildings* (NRC, 1990) and *Stewardship of Federal Facilities: A Proactive Strategy for Protecting the Nation's Public Assets* (NRC, 1998), both of which focused on the fiduciary responsibility of maintaining the nation's public assets. *Investments in Federal Facilities: Asset Management Strategies for the 21st Century* (NRC, 2004a) introduced strategies for investing in federal facilities from planning through disposal that were based on an analysis of best practices in private-sector and public-sector organizations. *Core Competencies for Federal Facilities Asset Management Through 2020* (NRC, 2008) projected the skills and knowledge necessary to manage federal facilities now and into the future.

During its first three meetings, the committee focused on gathering additional information from representatives of federal, private-sector, and professional organizations. The committee was briefed by the chair of the FFC, which sponsored this report. The committee also requested presentations from representatives of IBM, General Motors, General Dynamics, and the Association of Higher Education Facilities Officers-APPA and by three major providers of facility assessment consulting services: Parsons, Whitestone Research, and VFA Inc. (formerly VanderWeil Facility Advisors). Those organizations were contacted because the committee members on the basis of their experience with and knowledge of the facilities management profession believed them to be industry leaders in effective maintenance and repair-related practices. The committee also gathered information from the following federal agencies:

- U.S. Army Corps of Engineers
- U.S. Army Engineer and Research Development Center, Construction Engineering Research Laboratory
- Naval Facilities Engineering Command
- Naval Facilities Engineering Service Center
- NOAA
- DOE, including the Office of Science, the National Nuclear Security Administration, and the Office of Engineering and Construction Management
- U.S. Air Force
- Bureau of Overseas Buildings Operations of the U.S. Department of State
- U.S. Coast Guard
- National Aeronautics and Space Administration
- Smithsonian Institution
- Architect of the Capitol
- OMB

- GAO
- FASAB

The committee acknowledges that the various organizations that provided information do not represent a scientific or random sampling of organizations. Nor were the various organizations asked to comment on each other's processes and practices or to be involved in any way in the formulation of the committee's findings and recommendations.

In addition to the briefings, individual committee members researched the literature on international best practices for facilities management and finance, on risk and probability analysis models, and on tools and technologies for facilities management-related applications. The committee debated the issues and developed draft findings, then reviewed and integrated information to arrive at its final findings and recommendations. The resulting report represents a consensus of the committee that is based on a synthesis of the committee's data gathering and research, and on the individual members' expertise and experience. Statements based solely on the committee's collective opinions are so identified.

ORGANIZATION OF THIS REPORT

This report proposes new approaches to making decisions about allocating limited resources to achieve multiple benefits from investments in maintenance and repair of federal facilities and about managing the risks associated with deteriorating systems and components. It is addressed to several different audiences: federal facilities program managers, operating groups and their contractors; decision-makers in the administration and Congress; federal departments, agencies, and their advisors; and program and budget analysts throughout the federal government. Decision-makers, facilities program managers, and program and budget analysts in state and local governments and in other organizations may also find value in the report inasmuch as they face many of the same issues as their federal counterparts. Because this report addresses multiple audiences, different readers will find different chapters to be of greatest interest.

The Summary describes issues that are driving the need for a new approach to investments in maintenance and repair of federal facilities, identifies the basis of change, and contains the report's findings and recommendations. A more extensive discussion of the findings and recommendations is in Chapter 6.

Chapter 1, "Introduction," contains statistical information about the portfolio of federal facilities, long-standing facilities investment and management issues, the impetus for and foundation of more sustainable maintenance and repair practices, the committee's statement of task, and the committee's approach to fulfilling its task.

Chapter 2, "Outcomes and Risks Associated with Investments in Maintenance and Repair," briefly describes the processes used by federal agencies for identifying maintenance and repair requirements and for developing funding requests, the beneficial outcomes that can be expected from maintenance and repair investment, and the risks posed by deteriorating buildings, infrastructure, systems, and components.

Chapter 3, "Data, Tools, and Technologies to Support Investments in Maintenance and Repair," identifies an array of available and emerging technologies for data acquisition and tracking, indexes and models for measuring outcomes of investments in maintenance and repair, and predictive models for risk-based decision support.

Chapter 4, "Effective Practices for Investments in Maintenance and Repair," describes strategic practices used by public-sector and private-sector organizations that would improve the outcomes of investments in maintenance and repair of federal facilities.

Chapter 5, "Communicating Outcomes and Risk," focuses on strategies that federal facilities program managers can use to communicate more effectively with other decision-makers about the outcomes and risks associated with a given level of investment in facilities maintenance and repair.

Chapter 6, "Findings and Recommendations," presents detailed information about the committee's findings and restates its recommendations.

Chapter 7, "Implementing a Risk-Based Strategy for Investments in Federal Facilities' Maintenance and Repair," is intended to show how federal facilities program managers can put some of the report's recommendations into action. It suggests ways to quantify the beneficial outcomes, and offers guidelines for developing a longer-range strategic plan, guidelines for developing an annual budget submission, and methods for identifying risks related to deteriorating facilities' systems and components.

Appendixes A and B contain background information about the committee members and a list of committee meetings and briefings, respectively.

Appendix C is more technical than other sections of the report and is intended primarily for federal facilities managers. It provides some fundamentals of a risk-based approach, including basic principles of probability analysis, and examples of quantifying outcomes of maintenance and repair investments.

2

Outcomes and Risks Associated with Investments in Maintenance and Repair

Buildings, structures, and infrastructure pass through a number of stages during their lifetimes: planning, design, construction, operations and maintenance, renewal and revitalization, and disposal (Figure 2.1). The total cost of ownership of facilities is the total of all expenditures that an owner will make over all of those phases (NRC, 2008).

However, the amounts and distributions of those expenditures are not equal. Design and construction which require large capital expenditures but will typically last fewer than 5 years account for 5 to 10 percent of the total cost of ownership. In contrast, the operations and maintenance of facilities will require annual expenditures for 30 or more years and will account for as much as 80 percent of the total cost of ownership (NRC, 1998).

Buildings, structures, and infrastructure are composed of many separate but interrelated systems, including roofs, walls, windows, doors, cladding materials, foundations, mechanical, electrical, plumbing, heating, cooling, ventilation, communications, control systems, information technologies, security, fire, and safety. Each type of system is composed of individual components (such as valves, switches, coils, drainage pans, and materials), all of which must be kept in good working order if facilities are to perform as they were designed to do. A facility's overall performance is a function of the interactions of those systems and components, of interactions with the occupants, of the original design, and of operations and maintenance procedures.

How long facilities' systems and components actually perform at a satisfactory level (service lives) depends on many factors, including the quality of the original design, the durability of materials, the incorporated technology, location and climate, use and intensity of use, and the amount and timing of investment in

27

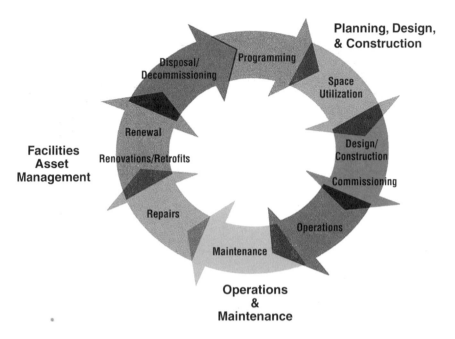

FIGURE 2.1 Facilities life-cycle model. SOURCE: NRC, 2008.

maintenance and repair activities. The service lives of systems and components can be optimized or at least improved by timely and adequate maintenance and repairs. Conversely, when maintenance and repair investments are not made when they are needed, the service lives of systems and components will be shortened (Figure 2.2).

TYPICAL OUTCOMES OF INVESTMENTS
IN MAINTENANCE AND REPAIR

A typical maintenance and repair program includes several types of activities that address different aspects or components of facilities' systems and have different objectives and outcomes. Maintenance is typically a continuous activity that addresses routine work that is accomplished on a recurring basis and includes some minor repairs. More important and often more expensive repair requirements are typically identified as separate projects. When federal facilities managers identify specific maintenance and repair requirements in funding requests, the funding for maintenance activities is typically presented as one lump sum and individual repair projects above some dollar threshold are identified separately. Projects that are identified as required but not funded make up the bulk of the backlog of deferred maintenance and repair projects.

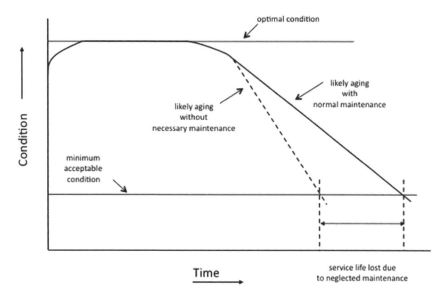

FIGURE 2.2 Effect of adequate and timely maintenance and repairs on the service life of a facility. SOURCE: NRC, 1993, 1998.

Maintenance and repair activities include the following (FFC, 1996):

- Preventive maintenance, which includes planned, scheduled, periodic inspection, adjustment, cleaning, lubrication, parts replacement, and minor repair of equipment and systems.
- Programmed major maintenance, which includes maintenance tasks whose cycle exceeds 1 year (such as painting, roof maintenance, road and parking lot maintenance, and utility system maintenance).
- Predictive testing and inspection activities that involve the use of technologies to monitor the condition of systems and equipment and to predict their failure.
- Routine repairs to restore a system or piece of equipment to its original capacity, efficiency, or capability.
- Emergency service calls or requests for system or equipment repairs that—unlike preventive maintenance work—are unscheduled and unanticipated.

All of these activities and projects are intended to do one or more of the following:

- Prevent a situation, breakdown, or failure that could result in unplanned outages and downtime that disrupt the delivery of programs or operations undertaken in support of organizational missions or that could result in the loss of life, property, artifacts, or research.

- Comply with legal regulations for safety and health.
- Extend the service life of building systems and components.
- Upgrade the condition of building systems and components to bring them back to original operating performance.
- Avoid higher future costs through timely investment and efficient operations.
- Respond to stakeholder requests.

Because of the interrelated nature of the systems and components embedded in facilities, maintenance or repair of one system or component can result in improvements in others. For that reason, investments in maintenance and repair can result in multiple outcomes that achieve several purposes. Maintenance and repairs that reduce energy and water use, for example, will also lower operating costs and provide for more efficient operations.

The beneficial outcomes that can result from maintenance and repair investments are shown in Table 2.1 and described below. They are grouped by their primary purposes in recognition that an outcome can be related to more than one purpose.

TABLE 2.1 Beneficial Outcomes Related to Investments in Maintenance and Repair

Mission-Related Outcomes	Compliance-Related Outcomes	Condition-Related Outcomes	Efficient Operations	Stakeholder-Driven Outcomes
Improved reliability	Fewer accidents and injuries	Improved condition	Less reactive, unplanned maintenance and repair	Customer satisfaction
Improved productivity	Fewer building-related illnesses	Reduced backlog of deferred maintenance and repairs	Lower operating costs	Improved public image
Functionality	Fewer insurance claims, lawsuits, and regulatory violations		Lower life-cycle costs	
Efficient space utilization			Cost avoidance	
			Reduced energy use	
			Reduced water use	
			Reduced greenhouse gas emissions	

Mission-Related Outcomes

Improved Reliability. Federal agencies require reliable supplies of power, heating, ventilation, air-conditioning, water, and other services to conduct their programs and achieve their missions. In such facilities as hospitals, research laboratories, museums, and military headquarters, those services are required 24 hours per day, 365 days per year to keep people safe and comfortable, to power equipment and computers, to ensure the integrity of research experiments, and to provide the constant temperature and humidity needed to protect cultural and historical artifacts, and works of art. Maintenance and repair activities are undertaken to ensure that mechanical, electrical, heating, ventilation, air-conditioning, and other systems are reliable and can perform without substantial interruptions, so that agencies can operate continuously on a routine basis, and during and after military operations, natural disasters, or manmade crises.

Improved Productivity. Maintenance and repair activities that support reliability also support improved productivity. Productivity for an individual or an organization has been defined as the quantity and/or the quality of the product or service delivered (Boyce et al., 2003). Productivity is most easily measured in manufacturing or similar functions where some number of units (such as cars or computer chips) with a given value can be expected to be produced per hour. If production goals fail to be met because of equipment or mechanical downtime, it is relatively easy to assign a dollar value to the effect. For example, the number of units that are not produced because of downtime can be multiplied by the sales value or the profit margin to arrive at a dollar value of lost productivity.

Productivity is less easily quantified for people engaged in administrative tasks, research, policy development, or many other tasks performed by federal employees, although it can be done in some situations. For example, the U.S. Patent and Trademark Office measures productivity by the number of patent applications reviewed per week (Campbell, 2011).

Functionality. Functionality is an assessment of how well a facility functions in support of an organizational mission. It also addresses a facility's capacity to meet the needs of occupants to navigate space and carry out activities (NIBS, 2008). Functionality loss, which is independent of condition, results from technical obsolescence, changes in user requirements, and changes in laws, regulations and policies. Thus, a facility can be in good condition but inadequate for its function. For example, a laboratory built in the 1970s may be well-maintained but still be technologically obsolete in light of today's research practices. Similarly, a facility that is in an earthquake zone but does not meet current seismic standards will be obsolete with regard to safety. Obsolete facilities that are in use not only fail to support organizational missions adequately, but siphon off resources for maintenance and repair. In some cases, it is more cost-effective to demolish an obsolete facility and replace it with one that is state-of-the-art, than to renovate and continue to operate it (NRC, 1993).

Efficient Space Utilization. The amount and type of space required by an agency to support its programs can be affected by changing missions, by such new work practices as telework, and by changing space standards. Vacant space or underused capacity within an occupied facility requires operating and maintenance resources as though the facility was fully utilized and siphons off the resources available for maintenance and repair activities of mission-critical facilities. Efficient use of space, in contrast, supports more cost-effective investment practices for maintenance and repair.

Compliance-Related Outcomes

Federal agencies, like other organizations, must comply with an array of safety and health regulations or face penalties for not doing so. Those regulations are intended to protect the health, safety, and welfare of workers and the public. They include regulations related to accessibility for people who have disabilities, potable water quality, occupational safety and health, and life-safety codes for fire suppression. Maintenance and repair activities that are undertaken to comply with regulatory standards include the replacement of obsolete, worn out or leaking plumbing components to bring them up to current standards and codes, the installation or modification of equipment to support accessibility for workers and members of the public who have disabilities, and preventive maintenance and testing of fire suppression and other life-safety systems.

Fewer Accidents and Injuries. Maintenance and repair investments are made to protect the safety of building occupants and visitors by eliminating hazards that can lead to accidents and injuries. Inadequate or dim lighting in buildings and stairways, torn carpeting and other hazards, can cause slips, trips, and falls that result in work-related injuries. Quality of lighting is also a factor in providing security and crime prevention in the workplace. Projects to upgrade floor coverings and provide slip-retardant surfaces or to provide better lighting can prevent accidents and injuries for workers and the visiting public. Projects to bring facilities up to current seismic codes can reduce the loss of life and property and reduce injuries if an earthquake occurs.

Fewer Building-Related Illnesses. The quality of indoor environments—concentrations of indoor contaminants such as chemicals and bioaerosols, temperature and humidity, lighting, ventilation, and noise levels—can influence a person's health, comfort, and ability to perform his or her job productively. Building-related illnesses and symptoms are substantially preventable through timely intervention to limit or eliminate exposure to causal agents, appropriate building design and construction, and good maintenance, operations, and cleaning practices (FFC, 2005). Maintenance and repair activities that can help to prevent building-related illnesses include the prevention of water intrusion that can result in indoor dampness and mold, regular replacement of filters in equipment, cleaning of coil drainage pans, and removal or encapsulation of asbestos.

Fewer Insurance Claims, Lawsuits, and Regulatory Violations. Preventing building-related accidents, injuries and illnesses, and complying with regulations can also result in fewer insurance claims and lawsuits, and fewer violations of health and safety regulations, and all their associated costs.

Condition-Related Outcomes

Improved Condition. Condition refers to the state of a facility with regard to appearance, quality, and performance. Investments to improve the condition of facilities, particularly in respect to the efficient performance of systems and components, often result in multiple beneficial outcomes.

Reduced Backlogs of Deferred Maintenance and Repair. This is an amount, expressed in terms of dollars, of the total deferred maintenance and repair work necessary to bring facilities back to their original designed performance capability, including updates required to meet current building and life-safety codes.

The importance of the existence of deferred maintenance is that it "implies that the quality and/or reliability of service provided by infrastructure on which maintenance has been deferred is lower than it should be, and thus the infrastructure is not or will not later be adequately servicing the public" (Urban Institute, 1994, p. 1). Another report found that "in the short-term deferring maintenance will diminish the quality of building services. In the long-term, deferred maintenance can lead to shortened building life and reduced asset value" (APWA, 1992, p. 1). As noted in Chapter 1, increasing backlogs of maintenance and repair projects create a fiscal exposure for the government which, in turn, affects the government's fiscal soundness.

Outcomes Related to Efficient Operations

Less Reactive, Unplanned Maintenance and Repair. A facilities management organization is more efficient when maintenance and repair activities are planned and scheduled not only to prolong the service lives of existing components and equipment but to replace them before a breakdown results in adverse events. Manpower is wasted when a large percentage of staff time is spent in reacting to unexpected breakdowns and through lack of planning that fails to incorporate potential efficiencies.

Lower Operating Costs. Operating costs include such elements as energy and water use, custodial services, security, fire suppression and detection, alarm testing and servicing, and grounds care. Timely maintenance and repair investments to ensure that heating, ventilation, and air-conditioning (HVAC) systems are operating properly can reduce energy use and costs and improve indoor environmental quality. Similarly, efficient plumbing systems can reduce the use and costs of water, and efficient fire systems can reduce false alarms, testing costs, and lost productivity due to unnecessary building evacuations.

Lower Life-Cycle Costs. In some cases, a modest investment in maintenance and repair can result in longer-term cost savings or extended service life. For example, replacement of a low-efficiency heat pump could pay for itself in a few years through savings in energy costs. Similarly, regular floodcoating of roofs could extend their service lives and delay the need to invest in replacement roofs.

Cost Avoidance. Cost avoidance results from making investments in the near term that avoid making larger investments later—a key objective of preventive maintenance activities. Examples include lubricating equipment components to avoid replacing the entire system, fixing minor roof leaks to avoid total roof replacement, applying protective coatings to avoid replacing the siding on a building or to avoid replacing equipment because of corrosion, and realigning equipment periodically to avoid shortening of service life due to wear and tear. Timely maintenance and repair can also avoid the need to keep large inventories of spare parts on hand and avoid unplanned service calls.

Reductions in Energy Use, Water Use, and in Greenhouse Gas Emissions. Maintenance and repair activities—such as replacement of malfunctioning cooling systems, replacement of lighting-system components with more efficient ones, and replacement of worn out roofs with "cool" roofs[1]—can result in reductions in energy use, water use, and greenhouse gas emissions. These actions often can help agencies to meet various mandates for high-performance facilities and related public policy objectives.

Stakeholder-Driven Outcomes

Stakeholders in maintenance and repair investments include not only facility managers, users, and tenants, but also the OMB, which is responsible for investment oversight; Congress; the administration; and the public. Each group of stakeholders has different expectations for the outcomes that should be achieved through investments in maintenance and repair of federal facilities (Figure 2.3).

Customer Satisfaction. Customer satisfaction as used in this report is an outcome related to the quality of services provided to facilities' users and tenants. Continuous and efficient operations of systems helps to create productive, safe, and healthy indoor environments. Conversely, failure of systems causes work disruptions, and inefficient operations or poor maintenance may result in poor indoor air quality and other adverse effects. Customer service calls related to temperature (too hot or too cold), humidity levels, moisture intrusion, air quality (odors), lack of ventilation, and water quality (tastes bad) can indicate that systems are not operating properly and require maintenance, repair, or replacement.

Improved Public Image. The appearance and upkeep of federal buildings can create a favorable or unfavorable impression for all stakeholders. Maintain-

---

[1]Cool roofs include white roofs, which stay cooler in the sun by reflecting incident sunlight back into space, and green (vegetative) roofs, which absorb rainwater and then cool by evapotranspiration.

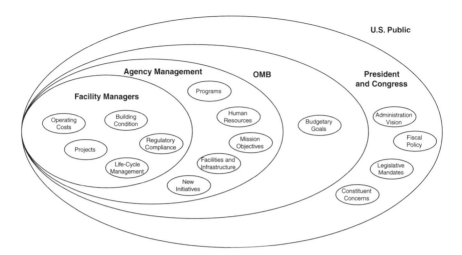

FIGURE 2.3 Stakeholders in federal facilities investments. SOURCE: NRC, 2004a.

ing the physical appearance and user accessibility of such iconic buildings as the U.S. Capitol, the White House, and the Washington Monument, is important for the national image of the United States in the eyes of its citizens and visitors. The upkeep and appearance of national park facilities, national museums, archives, and other facilities regularly visited by the public are important for visitors' experience and for their perception of how wisely tax money is being spent.

RISKS POSED BY DETERIORATING FACILITIES

The beneficial outcomes that can result from maintenance and repair investments are related not only to the total resources invested, but to how those resources are invested. Because the demands for resources for all federal programs will exceed available resources in coming years, priorities will need to be established for investments, and tradeoffs will need to be made. Risk assessment is an important tool for decision-making in a resource-constrained operating environment.

Risk-assessment processes have been used by federal, state, and local government agencies, by industry, and by academia for many years and for many applications. Organizations typically use risk assessment to inform themselves and the public about hazards presented by food, drugs, toys, air and water quality, and terrorism, and about the different actions or policy options that are available to manage the risks (NRC, 2009).

The essence of risk assessment as applied to facilities and building system components is captured by the three questions posed originally for risk assessment of nuclear reactors by Kaplan and Garrick (1981):

1. What can go wrong?
2. What are the chances that something with serious consequences will go wrong?
3. What are the consequences if something does go wrong?

The equivalent questions for risk management were posed later by Haimes (1991):

4. What can be done and what options are available?
5. What are the associated tradeoffs in terms of all costs, benefits, and risks?
6. What are the impacts of current management decisions on future options?

More recently, Greenberg (2009) framed the risk management questions, as follows:

4. How can the consequences be prevented or reduced?
5. How can recovery be enhanced if the scenario occurs?
6. How can key local officials, expert staff, and the public be informed to reduce concern and increase trust and confidence?

Just as maintenance and repair investments can result in an array of beneficial outcomes, the lack of investment and the deferral of needed maintenance and repair projects can result in adverse events (what can go wrong). Adverse events include more interruptions or stoppages of operations, more accidents, injuries, and illnesses, more lawsuits and insurance claims, increased operating costs, shortened service lives of equipment and components, failure to meet public policy objectives, and damage to the federal government's public image.

Risk—a measure of the probability and severity of adverse effects—will increase as federal facilities, building systems, and components continue to deteriorate through wear and tear and lack of investment. (The likelihood of an event occurring and of its consequences is also related to geography, climate, and other factors.) The risks associated with deteriorating facilities and systems identified by the committee are the following:

- Risk to federal agencies' missions. The risks related to lack of reliability including unplanned interruptions and downtime of facilities' systems and components; related to the diversion of resources to excess, obsolete, and underutilized facilities; and related to lowered productivity.
- Risk to safe, healthy, and secure workplaces. The risks related to increased injuries, illnesses, or even deaths involving federal personnel, contractor personnel, and the public; related to more lawsuits and claims resulting from facilities-related hazards; related to poor indoor environmental quality; and related to failure to comply with regulations.

- Risk to the government's fiscal soundness and public image. The risks related to the ownership of excess, underutilized, and deteriorating buildings; related to growing backlogs of deferred maintenance and repair projects; and related to higher operating and life-cycle costs.
- Risk to efficient operations. The risks related to underperforming facilities that drive up agency operating costs; related to customer dissatisfaction; and related to practices that fail to result in cost avoidances and other operational efficiencies.
- Risk to achieving public policy objectives. The risks related to the excessive use of energy, water, and other natural resources and to the production of greenhouse gas emissions.

3

Data, Tools, and Technologies to Support Investments in Maintenance and Repair

Reliable and appropriate data and information are essential for measuring and predicting beneficial outcomes of investments in maintenance and repair and for predicting the adverse outcomes of lack of investment. Data and information can be the basis of higher situational awareness during decision-making, of transparency during the planning and execution of maintenance and repair activities, of an understanding of the consequences of alternative investment strategies, and of increased accountability.

A 2004 National Research Council study stated that to implement a portfolio-based facilities asset management program effectively, the following elements are required (NRC, 2004a):

- Accurate data for the entire facilities portfolio to enable life-cycle decision making.
- Models for predicting the condition and performance of the portfolio of facilities.
- Engineering and economic decision-support tools for analyzing tradeoffs among competing investment approaches.
- Performance measures to evaluate the effects of different types of actions (such as maintenance versus renewal) and to evaluate the timing of investments.

This chapter focuses on the data, tools, and technologies that can be used to support portfolio-based facilities management and to support more strategic decision-making about investments in maintenance and repair. It is organized by data acquisition and tracking systems, indexes and models for measuring outcomes, and predictive models for decision support (Table 3.1).

The costs associated with data collection, analysis, and maintenance can be substantial. Costs will depend on the amount and accuracy of the data collected, how often they are collected, and the cost of the entire process, including data entry, storage, and staff time (NRC, 1998). Once data on a facility are created, it is necessary to update them throughout the facility's life cycle. The types of data to be maintained, their level of detail, and their currency, integrity, and attributes will depend on the outcomes that they are related to and how important the outcomes are for strategic decision-making. The data on facilities or systems that are mission-critical, for example, might need to be updated more often than data on less strategic facilities.

Because of the costs, the committee believes that "no data before their time" should be an infrastructure-management tenet. Every system and data item should be directly related to decision-making at some level, and off-the-shelf decision-support systems should be fully integrated into decision-making processes. To the greatest extent possible, data should be collected in a uniform manner across federal agencies to provide greater uniformity and in turn support the development of governmentwide performance measures and the greater use of benchmarking for agency practices and investment strategies.

DATA ACQUISITION AND TRACKING SYSTEMS

Facilities asset management data should include at least inventory data (number, locations, types, and size of facilities) that are relatively static once collected, and attribute data or characteristics that change (for example, equipment and systems, condition, space utilization, tenants, maintenance history, value, and age) (NRC, 2004a). The systems described below are designed to assist facilities managers to gather and maintain accurate, relevant data about an individual building or structure throughout its life cycle. Some are traditional passive systems that rely primarily on manual entry of data and others collect data automatically in "real time."

Traditional Passive Facilities Data-Acquisition Systems

Among the systems most commonly used by federal agencies for collection of data on portfolios of facilities are the following:

• **Computer-Aided Facility Management Systems.** Computer-aided facility management (CAFM) systems have evolved over several decades and through several generations of technology. However, from the beginning, the primary focus of such systems has been space planning and management and asset management. Applications now include energy and lease management, real

TABLE 3.1 Data, Tools, and Technologies to Support Strategic Decision-Making for Investments in Facilities Maintenance and Repair

Data, Tools, and Technologies	Primary Purpose and Description
Data Acquisition and Tracking Systems	
Computer-aided facility management systems (CAFM)	Space planning and management; facilities management
Computerized maintenance management systems (CMMS)	Maintenance-related work management
Building automation systems	Monitoring and control of lighting, heating, ventilation, air-conditioning and other building systems
Bar codes	Tracking of equipment, components, or other assets
Radio frequency identification systems (RFID)	Real-time asset tracking
Sensors	Monitoring of equipment and systems for vibration, strain, energy use, temperature, presence of hazardous materials, and the like
Condition assessments	Assessment of the physical condition of facilities systems and components
Hand-held devices	Allowing facility inspectors to enter work-management, condition, and other information directly into CAFM and CMMS
Automated inspections	Inspection of infrastructure, such as roads and railroads, using an array of technologies
Nondestructive testing	Monitoring of the condition of systems and infrastructure that are not visible to the human eye
Self-configuring systems	Control and other building systems, such as HVAC, that are able to diagnose a problem and fix it with minimal human intervention
Machine vision	An emerging technology for conducting inspections and for developing as-built information to support building information modeling
Building information modeling (BIM)	An emerging practice for modeling and exchanging of physical, financial, and other facility-related information throughout a facility's life cycle
Indexes and Models for Measuring Outcomes	
Facility condition index (FCI)	A financial index based on a ratio of backlog of maintenance and repair to plant value or current replacement value
Condition index	A financial index based on a ratio of repair needs to plant value or current replacement value

TABLE 3.1 Continued

Data, Tools, and Technologies	Primary Purpose and Description
Engineering-research-based condition indexes	Physical condition indexes based on empirical engineering research and developed from models; indexes have been developed for buildings, some building components (such as roofs) and a variety of infrastructure, including railroad tracks, airfield pavements, and roads
Building functionality index	An index to measure building functionality in relation to 14 categories; functionality requirements are independent of condition and are generally related to user requirements (mission), technological efficiency or obsolescence, and regulatory and code compliance
Building performance index	An index based on the ratio of a physical building condition index and a building functionality index

Predictive Models for Decision Support

Service life and remaining service life models	Models to predict the expected service life or remaining service life of systems and components; the purpose of these models is to help determine the appropriate timing of investments in maintenance and repair or replacement
Weibull models	Models that estimate the probability of failure of building or infrastructure systems or components
Engineering analysis	Analyses (such as fatigue analysis and wear-rate analysis) used to predict the remaining life of a system or component
Parametric models for cost estimating or budgeting	Economic-based (such as depreciation) or engineering-based (such as physical condition) models that can be used to develop multiyear maintenance and repair programs and cost estimates for annual budget development
Operations research models	An array of decision-support models that have been applied to some types of infrastructure (such as bridges)
Simulation models	Models used to analyze the results of "what if?" scenarios; an example is the Integrated Multiyear Prioritization and Analysis Tool (IMPACT), which simulates the annual fiscal cycle of work planning and execution; it can be used to set priorities for maintenance and repair work based on different variables, including budget
Proprietary models	Facilities asset models developed for a wide array of applications, including the prediction of outcomes of investments for maintenance and repair developed by private-sector organizations; relatively little information is publically available about how they work and their assumptions, robustness or accuracy

estate management, maintenance and operations, and geographic information systems (GIS) integration. The latest incarnation of CAFM systems, integrated workplace management systems (IWMS), emphasize the integration of all those applications with an organization's financial and human resources data systems. All aspects of a facility's life cycle—including planning, design, financial analysis and management, project management, operations, facilities management, and disposal—are accounted for in IWMS.

• **Computerized Maintenance Management Systems**. Computerized maintenance management systems (CMMS) have also evolved, with maintenance-related work management as their primary focus. Today's fully developed CMMS can be used for preventive maintenance scheduling, labor requirements, work-order management, material and inventory management, and vendor management. Most CMMS track facilities-related materials, location, criticality, warranty information, maintenance history, cost and condition, component assembly, and safety information.

One drawback of CMMS and CAFM is that data often are entered manually and this increases the likelihood of error. In addition, the information available to decision-makers can be compromised in that manually updated data may not be entered on a timely basis. However, if the data are accurate, complete, and current, such systems provide a database that can be used for measuring outcomes, for risk analysis, for energy and condition assessment modeling, and for investment-decision support.

Real-Time Active Facility-Data Acquisition Systems

Several technologies bring "intelligence" to facilities, systems, and components; allow for automated data entry into CMMS and CAFM systems; and allow real-time monitoring of the performance of facilities systems and components.

• **Building Automation Systems.** Building automation systems (BAS) are typically installed to monitor and control lighting, heating, ventilation, and air-conditioning (HVAC) systems; security systems; and life-safety systems, such as fire suppression. These control systems provide real-time feedback in the form of alarms based on operating characteristics (an alarm sounds when specified parameters are exceeded) and records of equipment performance. They are often the best source for early detection of equipment problems.

Traditionally, the data collected by BAS were used only for control purposes. Today, BAS data are being "mined" to provide information so that system operators and facilities managers can understand and assess the performance and condition of systems. For example, advanced sensors can be installed in BAS for HVAC systems (Liu and Akinci, 2009) to track temperature, humidity, pressure, and ventilation rates (ASHRAE, 2009), all of which are related to indoor environmental quality. The data collected can be used to support more effective decision-making about the daily operations of facilities and their systems (Jones and Bukowski, 2001; Du and Jin, 2007) and to support the health, safety, and productivity of building occupants.

The use of open protocols—open standards by which devices communicate with each other—makes it possible to connect and control devices from multiple BAS developed by multiple vendors. With the arrival of the Internet and Internet Protocol and wireless technology, the performance of systems can now, at least theoretically, be monitored and controlled no matter where they are and no matter who manufactured them.

The reality, however, is that many BAS were installed at different times, by different vendors, and in connection with different generations and types of systems, and this has resulted in a lack of interoperability among hardware, software, and communication protocols. Nevertheless, there are no technologic hurdles that need to be overcome to enable BAS to be used for portfolio-based facilities management and resource planning, or to allow coordinated operations of building equipment and systems to achieve high operating efficiency, while minimizing operating times, and reducing energy use.

• **Bar Codes.** Bar codes involve an older technology that has evolved and become more complex with greater capabilities. When bar codes are placed on equipment, components, or other assets, information about assets can be automatically scanned into a CMMS or CAFM system.

• **Radio Frequency Identification Systems.** Radio frequency identification (RFID) systems use two technologies: a radio tag that consists of a microchip that stores data on an object and an antenna that transmits data; and a reader that creates the power for the microchip in the tag for passive RFID tags and then reads and processes the data from the tag. The radiowave data transmitted by the tag are translated into digital information that in turn, can be used by the software to record the status or location of a facility, a system, or another component.

RFID tags can replace bar code systems for traditional inventory applications. Embedding a tab in an asset increases the durability and reusability of the tag. Using RFID tags with sensors substantially increases the number of applications for monitoring performance. Sensors are most often used to monitor motion (vibration) of and strain on facilities systems and components or to monitor temperature. Increasingly, such sensors (often in conjunction with BAS) are being used for energy management systems. Other sensors can be used to detect the presence of radiation, chemicals, or other hazardous materials.

RFID systems are only now beginning to be used for facilities management-related activities. Given their numerous advantages over traditional bar coding and their decreasing size and cost, it seems only a matter of time before they replace bar coding. Although there are stand-alone RFID systems that offer alerts, customized reports, and other features, the integration of RFID systems with real-time

BAS or passive CAFM, CMMS, and IWMS makes possible a greater variety of applications, more efficient use of data, and more efficient operations.[1]

Condition Assessment Data Collection and Tracking

Condition is an underlying factor in the performance of most facilities, systems, and components. It is also an important predictor of future performance: systems and components that are in good condition will be more reliable and perform better than systems that are deteriorating.

Condition assessments provide reference points for facilities managers on the current condition of facilities, systems, and components. Trends in condition can be used to determine whether facility systems and components are being maintained and are meeting their expected service lives or whether their performance is deteriorating faster than expected.

Information about the condition of facilities, their systems, and their components can be gathered and updated by using different approaches, whose costs vary. The choice of approach will affect the availability, timeliness, and accuracy of data and thus affect the value of the data for strategic decision-making.

In federal agencies, condition-related data are typically acquired through assessments conducted by teams of inspectors on a multiyear cycle. Condition assessments are usually conducted each year for a portion of an agency's entire portfolio and most facilities are inspected once every 3 to 5 years.

Depending on the information needed and how an organization uses the results, condition assessments can range from detailed assessments of individual components by engineers or technicians of various specialties to walk-through visual inspections by small teams. Condition assessments of any kind help to verify assumptions about facility system conditions and to update real estate records. The consistency and quality of condition assessments among facilities and sites are also important in determining the usefulness of the data collected for decision support and priority-setting.

The costs of condition assessments vary widely, depending on the complexity of the facility and the level of inspection. Cost estimates given in presentations to the committee ranged from $0.07 to $0.60 per square foot of building space when third-party contractors performed the initial condition assessment. For a 500,000 square foot facility, that would translate to $35,000 to $300,000.

Condition assessments that are undertaken on a multiyear cycle and conducted for an entire portfolio of facilities can be inefficient and expensive and the

---

[1]Bar coding is considered less expensive than RFID technology: it is estimated that bar codes cost $0.005 each whereas passive RFID tags cost more than $0.05 each (Shih, 2009). However, when other variables (such as the speed of collecting data or the cost of RFID scanners versus barcode interrogators and the number of times that an asset inventory is performed) are considered, the cost differentials become smaller (Roberti, 2009). If the inventory system uses multiple bar codes, it is usually a sign that RFID technology will be more cost-effective than traditional bar coding.

information can lose its value for decision-making quickly. Some organizations are now taking a "knowledge-based" approach to condition assessment. The term knowledge-based is used "to indicate that knowledge (quantifiable information) about a facility's system and component inventory is used to select the appropriate inspection type and schedule throughout a component's life cycle. Thus, inspections are planned and executed based on knowledge, not the calendar" (Uzarski et al., 2007, p. 2). Because different building components have different service lives, and some may be more important than others with respect to outcomes and risks, some components are inspected more often than others and at different levels of detail. By tailoring the frequency and level of inspections, a knowledge-based approach makes better use of the available resources and provides more timely and accurate data to support investment-related decisions (Uzarski, 2006).

• **Hand-Held Devices.** Hand-held devices and kiosk terminals placed throughout an organization's facilities allow inspectors and building operators to enter data on work orders or other building-related actions in nearly real time directly into CAFM and CMMS. That provides a more accurate and up-to-date picture of current maintenance efforts and requirements and of building condition. Hand-held devices, when used by properly trained maintenance staff, reduce the time spent in recording information, improve data accuracy, and allow more time to be dedicated to hands-on maintenance and repair activities.

• **Automated Inspections.** Technologies exist that replace the human inspector for gathering condition-related data. One example is the International Road Roughness Method, which is in wide use around the world by the highway industry (Gillespie et al., 1986). The railroad industry routinely uses laser optical sensors, accelerometers, displacement transducers, motion detectors, and gyroscopes for measuring track quality under a moving load, deviations from which increase derailment risk and adversely affect operations. The use of those types of technologies allows the collection of more data and higher-quality data at a fraction of the cost of manual data collection (Union Pacific, 2005).

• **Nondestructive Testing.** Nondestructive testing is sometimes used for collecting condition-related data on components and systems not visible to the human eye. Many technologies can be used for these purposes. Among them are infrared thermography for detecting excessive heat, leaks, delamination, and defective areas and for stress mapping; ultrasonic testing and laser technology for detecting cracks and other defects; and ground-penetrating radar for detecting abnormalities in subsurface systems (NRC, 1998).

Emerging Technologies for Data Acquisition and Tracking

Technologies for various aspects of facilities management are continually evolving and advancing in their capabilities. Three technologies that could substantially improve the acquisition and tracking of data, improve maintenance and repair activities, and provide support for decision-making are self-configuring systems, machine vision, and building information modeling (BIM). These technologies are described below.

• **Self-Configuring Systems.** A self-configuring (or self-healing) system is one that is capable of responding to changing contexts in such a way that it achieves a target behavior by regulating itself (Williams and Nayak, 1996). The objective of self-configuring systems is to enable computer systems and applications to manage themselves with minimal but high-level guidance by humans (Parashar and Hariri, 2005).

For example, self-configuring HVAC systems detect and diagnose a problem (such as a damper stuck in a variable-air-volume box) and automatically fix it (Laster and Olatunji, 2007). With continuous monitoring, such systems could recognize that they are going out of commission and then repair themselves, or they could identify ways to reduce energy costs and improve occupancy comfort. The benefits of self-configuring or self-healing systems could include lower operating costs, greater reliability, less downtime, and more efficient operation (Fernandez et al., 2010). Some of those characteristics have been studied in the BAS domain (Ellis and Mathews, 2002; Sallans et al., 2006; Menzel and Pesch, 2008).

• **Machine Vision.** Machine vision is an emerging technology for conducting facility inspections. Machine vision uses video imaging and computer software to detect component defects, such as cracks. The technology has been used for pavement inspection (Tsai et al., 2010) and railroad-car structural inspection (Schlake et al., 2010) and is under development for railroad-track inspection at the University of Illinois (Resendiz et al., 2010).

In the federal sector, the General Services Administration is developing a set of guidelines for rapid collection of 3-D information by using 3-D imaging technologies (in particular laser scanners) for historical and facility-condition documentation and for collecting as-built information that can be used in the development of building information models.[2]

• **Building Information Modeling.** Building information modeling (BIM) is an emerging practice for modeling and exchanging facility information that involves various interoperable technologies and associated sets of processes (Eastman et al., 2008; Smith, 2007). It has the potential to contain and visually display data about physical elements (such as columns, beams, slabs, and walls), nonphysical concepts (such as zones), and the relationships between them. BIM could also provide information about nongeometric properties and attributes, such as material specifications needed for fabrication, material properties that depict behaviors under different contexts (such as thermal, acoustic, and light reflectance), cost, budget, and schedule or even information about parametric rules that depict the connections and distances between objects. Configured in

---

[2]Additional information about GSA's 3-D laser scanning effort is available at http://www.gsa.gov/portal/content/102282. Accessed on 03/31/2011.

that way, BIM would incorporate consistent, coordinated, and nonredundant data on a facility (Eastman et al., 2008).

The use of BIM for planning, designing, and constructing facilities is increasing throughout the architecture-engineering-construction industry. The benefits of using BIM for energy simulation, cost estimation, subcontractor coordination, and other applications have been documented. Some federal agencies have begun to require the use of BIM during design and construction (Brucker et al., 2010). A number of BIM guides and roadmaps for future development have been developed by, for example, the Department of Veterans Affairs[3], the state of Wisconsin (Wisconsin, 2009), Indiana University (2010), the National BIM standard effort,[4] and the Associated General Contractors of America (AGC, 2006).

To date, BIM has been applied to facilities operations and management only sparsely, although using it for building operations, maintenance, and management could yield substantial benefits and long-term cost savings (Wisconsin, 2009). For example, a U.S. Coast Guard facility planning case study recorded a 98 percent reduction in time and effort in producing and updating a facility management database when BIM was used (Eastman et al., 2008; Dempsey, 2009). A case study of the Sydney Australia Opera House identified numerous benefits of using BIM, including consistency in data, providing an integrated source of information for different software applications, and supporting queries for data mining (Ballesty et al., 2007). Additional case studies highlight how BIM can be used for building-systems commissioning, field operations, asset tracking, and energy monitoring (Jordani, 2010).

One of the greatest benefits of using BIM would be having real-time facility-related information in an integrated form that would enable facility operators and managers to have a more holistic understanding of what is happening throughout a facility's life cycle. BIM would reduce redundant data collection and data reentry and reduce the uncertainty associated with not having the right information when making investment decisions. More accurate, real-time information would bring greater transparency to facility operations, which would increase accountability. Finally, a by-product of BIM is advanced 3-D visualization capabilities, which can help in communicating facility investment requirements and the predicted outcomes of investments of different stakeholders.

Although the benefits of BIM for facilities management and operations are apparent, BIM technology in its current form is best categorized as an information repository. Improved data-exchange standards and software systems are needed to allow full interoperability of data from many systems. Interoperability, in turn, will allow more seamless integration of the data and functionalities needed to support strategic decision-making related to maintenance and repair investments and to document the outcomes.

---

[3]The Department of Veterans Affairs BIM guide is available at http://www.cfm.va.gov/til/bim/ BIMGuide/lifecycle.htm.

[4]Information available at http://www.buildingsmartalliance.org/index/php/nbims/about/.

Various federal agencies participate in and support a number of efforts to develop the data and exchange standards, protocols, standard definitions, and data items that are needed if BIM is to reach its full potential as a tool for portfolio-based facilities management. Some of the efforts, including the development of a national BIM standard, are being conducted under the auspices of the National Institute of Building Sciences and involve representatives of federal and private-sector organizations.

INDEXES AND MODELS FOR MEASURING OUTCOMES

An array of indexes derived from models have been developed to measure outcomes related to building and infrastructure condition, functionality, and performance.

Condition Indexes and Models

Facility Condition Index. The facility condition index (FCI) is a well-known and widely used condition index modeled from the ratio of two direct monetary measures: backlog of maintenance and repair (cost of deficiencies) and current replacement value (NACUBO, 1991). Typically when applied to a facility, that ratio ranges from 0 to 1, but it is sometimes multiplied by 100 to expand the range from 0 to 100.

Condition Index. The Federal Real Property Council defines the condition index (CI) as a general measure of a constructed asset's condition at a specific time. CI is calculated as the ratio of repair needs to plant-replacement value (PRV): CI = (1 − $ repair needs / $ PRV) × 100 (GSA, 2009). Repair needs represents the amount of money necessary to ensure that the constructed asset is restored to a condition substantially equivalent to the originally intended and designed capacity, efficiency, or capability. PRV is the cost of replacing an existing asset at today's standards (GSA, 2009). Like FCI, CI is a financial measure that is a proxy for physical condition.

Engineering-Research-Based Condition Indexes. Engineering-research-based indexes measure the physical condition of facilities, their systems, and their components. These types of indexes are based on empirical engineering research and are the driving engines for the sustainment management systems (SMS) (decision-support systems and asset-management systems) developed by the U.S. Army Corps of Engineers. The indexes can be applied to airfield pavements (Shahin et al., 1976; Shahin, 2005), roads and streets (Shahin and Kohn, 1979), railroad track (Uzarski et al., 1993), roofing (Shahin et al., 1987), and building components (Uzarski and Burley, 1997).

Each index follows a mathematical weighted-deduct-density model in which a physical condition-related starting point of 100 points is established. Some number of points is then deducted on the basis of the presence of various distress types (such as broken, cracked, or otherwise damaged systems or components), their severity (effect), and their density (extent). The deductive values were based on a consensus of many building operators, engineers, and other subject matter experts. Risk assessment and consequence are incorporated into the severity definitions and the actual deductions to be taken for each combination. For example, a "high" severity generally denotes health, life-safety, or structural integrity problems or mission impairment. Inspectors need only collect distress data and they do not make judgments concerning physical condition. The computed building condition index (BCI) will be plus or minus 5 points of the expert-group consensus with 95 percent confidence on a 0 to 100 scale.

Those indexes are computed at a facility hierarchy level typically associated with maintenance and repair activities (for example, logical pavement portion, logical roof portion, and air-handling unit). Logical management units are based on component type, material or equipment type, location, age, and other discerning factors. Maintenance and repair needs are correlated to the numerical BCI scale. In general, the lower the BCI value, the greater the risk of physical failure. Different BCI scale ranges (such as 86 to 100 or 71 to 85) signify the relative risk. The indexes can also be rolled up to determine the condition of a system, facility, entire portfolio, or portfolio subsets to support reporting and managerial requirements.

To maximize the usefulness of the indexes, condition standards need to be established, that is, the point at which the component condition drops below a minimum desired value whereby a mission is adversely affected or the risk of mission impairment becomes unacceptable and triggers a maintenance and repair requirement. The minimum value is a variable and depends on the facility, mission, risk tolerance, redundancy, occupancy, location, and other factors. Representatives of the U.S. Navy told the committee that they are working on creating those types of standards for their facilities portfolio. The committee notes that determining the minimum values is not a trivial matter for any organization and that additional research is needed to facilitate such a determination.

Building Functionality Index and Model

Functionality is a broad term that applies to an entire facility and its capacity to support an organization's programs and mission effectively. Functionality is related primarily to user requirements (mission), technical obsolescence, and regulatory and code compliance, and it is independent of condition. A building functionality index (BFI) for buildings and building functional areas (such as administration, laboratory, storage, and production) has been developed by the U.S. Army (Grussing et al., 2009). It follows the same form, format, and rating-scale development theory as engineering-research-based physical condition indexes. However, rather than accounting for distresses, functionality issues are considered with severity (effect) and how widespread the issue is. The numerical BFI scale

(0 to 100) is correlated to modernization needs. The model addresses 65 specific functionality issues, which are grouped into 14 general functionality categories, as shown in Table 3.2.

Building Performance Index and Model

A building performance index (BPI) has been developed that combines the BCI and the BFI into a measure of the overall quality of a building. The BPI is derived mathematically by taking the sum of $^2/_3$ of the lowest of the BCI and BFI values and $^1/_3$ of the higher of the two values. The $^2/_3$ to $^1/_3$ split was derived through regression analysis and is intended to serve as a measure of rehabilitation needs.

TABLE 3.2 Building Functionality Index Categories and Descriptions

Category	Description
Location	Suitability of building location to mission performance
Building size and configuration	Suitability of building or area size and layout to the mission
Structural adequacy	Ability of structure to support seismic, wind, snow, and mission-related loads
Access	Ability of building or area to support required entry, navigation, and egress
Americans with Disabilities Act (ADA)	Level of compliance with the ADA
Antiterrorism and force protection (AT/FP)	Compliance with AT/FP requirements
Building services	Suitability of power, plumbing, telecommunication, security, and fuel distribution
Comfort	Suitability of temperature, humidity, noise, and lighting for facility occupants
Efficiency and obsolescence	Energy efficiency, water conservation, and HVAC zoning issues
Environmental and life-safety	Asbestos abatement, lead paint, air quality, fire protection, and similar issues
Missing and improper components	Availability and suitability of components necessary to support the mission
Aesthetics	Suitability of interior and exterior building appearance
Maintainability	Ease of maintenance for operational equipment
Cultural resources	Historic significance and integrity issues that affect use and modernization

SOURCE: Grussing et al., 2009.

PREDICTIVE MODELS FOR DECISION SUPPORT

Outcomes of maintenance and repair investments are measurable (either directly or through a model) by taking before and after measurements to gauge the effects of the investment. However, predictive models are needed to estimate the outcomes before investment (or in the absence of investment, that is, the do-nothing case). The prediction can be compared with the measured postinvestment value to determine whether the expected outcome was realized. Such predictions are crucial for performing a consequence analysis of maintenance and repair alternatives.

Modeling approaches have been developed to predict the remaining service lives of facility systems and components and to estimate the probability of system failure, and they support risk-based decision making related to the timing of maintenance and repair investments. Cost and budget models are also available to support the development of multiyear maintenance and repair programs.

Models of Service Life and Remaining Service Life

Service life is the expected usable life of a component. At the end of the service life, replacement or major rehabilitation or overhaul is required. Remaining service life is the time from today to when the service life will be expended.

Service life is based on a number of factors, which may include manufacturer's test data, actual in-service data, and opinion based on experience. Service life typically is expressed as 5, 10, 15, 25, or 50 years. In reality, service life is in a range because of operating environments, the magnitude and timing of maintenance, use, abuse, and other factors.

Knowing the service life and the remaining service life of a component is important for making decisions about the timing of investments and for planning maintenance and repair work. Service-life and remaining-service-life models can also consider risk. If risk tolerance is high, maintenance and repair investments can be planned for the year in which the service life is expected to expire (or beyond). If risk tolerance is low, maintenance and repair may be planned to occur before the remaining service life expires. As risk tolerance decreases, maintenance and repair activities will be implemented sooner rather than later in relation to remaining service life.

Efforts have been made to determine the lives of facilities as a whole. Such organizations as the Bureau of Economic Analysis and Marshall and Swift have published facility service-life information. Facility service lives are generally based on economic depreciation rather than physical condition and performance. Such values are useful for planning overall facility recapitalization and modernization, for computing commercial tax liability, and for appraising value (Whitestone, 2001).

The BUILDER Sustainment Management System[5] calculates service life on

---

[5]Information available at http://www.erdc.usace.army.mil/pls/erdcpub/docs/erdc/images/ERDC_FS_Product_BUILDER.pdf.

the basis of the predicted component-section condition index (CSCI) (see discussion of Weibull models below) and a standard CSCI value denoting a physical condition whereby the component should be replaced. Remaining service life (RSL) is the difference between the current age and the predicted service life. RSL is adjusted on the basis of the revised predicted CSCI value resulting from the most recent condition survey inspection (Uzarski et al., 2007).

Remaining Maintenance Life Model. Remaining maintenance life (RML) measures the time remaining before maintenance and repair should be accomplished (Uzarski et al., 2007). With the CSCI and a minimum acceptable condition standard needed to support a mission fully, RML can be predicted in the same manner as RSL (Uzarski, 2004).

Weibull Models. Weibull models estimate the probability of failure and are widely used to estimate the weak link in a system. They have been used in the railroad industry for predicting defect formation in rails (Orringer, 1990). The presence of defects and the defect rate (defects/mile) are criteria for planning rail-defect testing and rail replacement. Because of public-safety concerns, the risk tolerance for defect-caused rail breaks that result in derailments is very low.

Weibull models have also been applied to predicting the engineering-based CSCI in buildings (Grussing et al., 2006). In recognition of a probability that a component section will fail faster or slower than expected, a Weibull model was used in the BUILDER SMS to predict a "current" CSCI. The CSCI was predicted to overcome the fact that building components are not all inspected at the same time and that years may pass between inspections. The prediction model provides real-time condition reporting, including the rollup CIs. It is also used to predict future CSCI and rollup CIs.

Because of the variance in actual versus predicted service life of systems and components, service life must be adjusted in accordance with actual condition data. With the Weibull-based CSCI prediction model, the adjusted service life, RSL, and RML can be computed.

Engineering Analysis. Traditional engineering analyses and models are also used to predict remaining component life. Examples include fatigue analysis, wear-rate analysis, and corrosion effects on structural strength. Typically, components are analyzed only on an as-needed basis. Computations of stress or strain coupled with material properties (such as strength and dimensions) and operating conditions are often used in a model for estimating an outcome.

Cost and Budget Models

Predicting outcomes of maintenance and repair investments requires estimates of costs or budget needs and consequences. The traditional approach involves using cost estimates developed for individual projects. The expected outcomes of a list of projects can be married to the costs of the projects and the outcome of a maintenance and repair investment can be predicted. However, this approach has several problems. First, when the overall program and cost are being developed with planned outcomes, the projects to support the program may or may not have already been developed and their costs estimated. Second, the estimated costs may have been developed 2 years previously or earlier and may need to be updated. Third, developing project level cost estimates can be expensive so it typically is not done unless there is high certainty that the project will be funded. Finally, the costing approach is not particularly sensitive to what-if consequence analyses. Any change would require the cost estimator to recompute the cost with the changes incorporated.

To overcome the issues related to project cost estimating, parametric cost or budget models have been developed. Parametric models use cost correlated to particular measures to provide a reasonable estimate that is sufficiently accurate for planning purposes. They may be economic-based (involving, for example, depreciation or average service life) or engineering-based (involving, for example, actual and predicted condition or adjusted service life). Detailed project estimates are completed later in the project planning and execution process, when there is greater certainty about the availability of funding.

Examples of parametric models include the facility sustainment model developed for the Department of Defense (Whitestone and Jacobs, 2001) and the associated recapitalization and operating cost models (Lufkin et al., 2005). Those models are economics-based rather than engineering-based. The BUILDER SMS system uses an engineering-based parametric cost model based on component replacement cost and the CSCI. In general, parametric cost models are particularly useful for developing multiyear maintenance and repair programs.

Operations Research Models. Operations research (OR) models have been applied to management of some types of infrastructure, such as bridges (Golabi, 1997). However, although OR models are well-suited to maintenance and repair investments, they are seldom applied. With OR techniques (there are many), an objective function (such as minimizing energy consumption) could be established subject to budget, labor, and other constraints. Multiple criteria could be considered, and an optimal mix of projects and a prediction of the outcomes could be identified.

On the basis of an agency's goals and needs (such as resource allocation, maintenance and repair scheduling, and logistics), OR models could be used for a variety of normative analyses. For example, in many situations in which inspection and collection of data are expensive, OR models could be used to find the optimal frequency of inspection or the need for collecting more data or samples. The key tradeoff in this type of analysis involves making a decision based on the available information versus collecting more data before making a decision.

Stochastic optimization models are used in situations in which decision-makers are faced with uncertainty and must determine whether to act now or to wait and see. There is uncertainty in future facility deterioration, budget levels, the effects of maintenance and repair actions, and so on. Stochastic optimization models evaluate managerial recourse, and this provides an opportunity to fix problems if worse-case scenarios occur.

Another class of OR model, Markov decision process (MDP), attempts to link facility condition and optimal long-term maintenance strategy. MDP models are used for management of networks of facility systems and components (such as pavements and bridges). The application of MDP models in facility asset management is limited, in that, for example, facility condition may result from different deterioration processes and require different remedial actions. Deterioration processes and required remedial actions are difficult to define in a model.

Simulation Models. Simulation models are used to analyze the results of what-if scenarios and can be used in conjunction with OR models. Although much research is needed for simulation (deterministic and probabilistic) modeling with regard to facility maintenance and repair funding and consequence analysis (pertaining to outcomes), the U.S. Army Engineer Research and Development Center, Construction Engineering Research Laboratory (ERDC-CERL) has developed the deterministic Integrated Multiyear Prioritization and Analysis Tool (IMPACT) simulation model.[6] The IMPACT model simulates the annual fiscal cycle of work planning and execution and displays building, system, and component condition indexes up to 10 years into the future. The model, in part, sets priorities for maintenance and repair work and their assumptions and assigns funding on the basis of such variables as policies, standards, and budgets.

Proprietary Models. The private sector has long been active in facility-asset management and has collected large amounts of facility data. Some companies have used the data to develop asset management models. Generally, the models purport to predict condition and budgets, and outcomes related to both. However, such models are proprietary, so little about how they work, or about their assumptions, robustness, and accuracy, is publically known or peer-reviewed.

---

[6]Information available at http://www.cecer.army.mil/td/tips/product/details.cfm?ID=738.

4

Effective Practices for Investment in Maintenance and Repair

In 2002, the National Research Council appointed the Committee on Business Strategies for Public Capital Investment (NRC, 2004a, p. x) to

> develop guidelines for making improved public investment decisions about facilities and supporting infrastructure, their maintenance, renewal, replacement, and decommissioning. As part of this task, the committee was asked to review and appraise current practices used to support facilities decision-making in both the private and public sectors and identify objectives, practices, and performance measures to help determine appropriate levels of investment.

The resulting report, *Investments in Federal Facilities: Asset Management Strategies for the 21st Century,* identified 10 principles/policies used by best-practice organizations in matters of facilities investment and management (NRC, 2004a). The report noted that despite the inherent differences between public and private-sector organizations regarding goals, missions, and operating procedures, some aspects of all the principles/policies could be adapted to the federal operating environment.

To gather information for the present report, the committee identified private-sector companies and professional organizations that it believed to be industry leaders in effective maintenance and repair practices, and it heard directly from four: IBM, General Motors (GM), General Dynamics, and the Association of Higher Education Facilities Officers-APPA. Information was also obtained from three major providers of facility assessment consulting services—Parsons, Whitestone Research, and VFA Inc.—and from numerous federal agencies, as noted in Chapter 1.

The 2004 National Research Council report found that best-practice organizations all did the following (NRC, 2004a, p. 2):

- Establish a framework of procedures, required information, and valuation criteria that aligns the goals, objectives, and values of their individual decision-making and operating groups to achieve the organization's overall mission; create an effective decision-making environment; and provide a basis for measuring and improving the outcomes of facilities investments. The components of the framework are understood and used by all leadership and management levels.

- Implement a systematic facilities asset management approach that allows for a broad-based understanding of the condition and functionality of their facilities portfolios—as distinct from their individual projects—in relation to their organizational missions. Best-practice organizations ensure that their facilities and infrastructure managers possess both the technical expertise and the financial analysis skills to implement a portfolio-based approach.

- Integrate facilities investment decisions into their organizational strategic planning processes. Best-practice organizations evaluate facilities investment proposals as mission enablers rather than solely as costs.

General Dynamics, IBM, and GM all follow those practices for managing their facilities. They also reported that they had been successful in obtaining adequate funding for their maintenance and repair programs. They attributed their success, in part, to a combination of strategies as follows:

- Facilities are closely aligned with the organization's mission—excess or underutilized facilities are disposed of and space is proactively managed to minimize the total square footage in use.
- Maintenance and repair investments are linked to the organization's product delivery or bottom line because failure to invest can result in financial harm to the organization and real or perceived harm to its capacity to perform.
- Investments are made to ensure compliance with regulatory or statutory requirements because failure to do so can result in legal and financial penalties.
- The work undertaken results in efficient operations, which result in lower operating costs that can be documented.

The private-sector representatives identified a number of practices that are used by their organizations to ensure that maintenance and repair investments result in outcomes that are beneficial to the entire organization:

- Dispose of excess and underutilized facilities (buildings, structures, and infrastructure).
- Pursue a proactive strategy to minimize the total facilities "footprint."

- Correlate the effects of failure with the organization's mission.
- Correlate repair delay with sustainment cost.
- Remove "must-fund" projects and those supported by acceptable financial payback from the maintenance and repair account and fund them with other discretionary sources.
- Use consistent standards to strategically assess the condition of facilities that require maintenance and repair.
- Conduct a year-end budget review to evaluate investment performance.

The committee acknowledges that the choice of those practices is not based on an industrywide survey of best-practice organizations or on a scientific or random sampling of organizations. Nonetheless, it believes that if such practices were implemented in federal agencies, they could result in more cost-effective practices that would yield improved long-term results, as described below.

DISPOSE OF EXCESS AND UNDERUTILIZED FACILITIES

Effective portfolio-based facilities management looks holistically at the entire inventory of buildings and structures and aligns them with the organization's overall mission and operating objectives. Continual monitoring is required to identify facilities that become excess or underutilized because of changes in requirements or in the operating environment.

Private-sector organizations "have a direct incentive to dispose of unneeded facilities because they are a drain on organizational resources and are readily identifiable on their balance sheets" (NRC, 2004a, p. 97). Excess or underutilized facilities are disposed of through sales, demolition, or nonrenewal or termination of leases to free resources for other organizational requirements. In that way, private-sector organizations manage the risk of fiscal exposure related to the ownership of facilities, reduce their maintenance and repair requirements, and reduce facilities-related expenses, such as property taxes, energy and water, insurance, and security. They also manage the risk to their public image posed by abandoned and poorly maintained facilities, which could affect the public's willingness to buy their products (NRC, 2004a).

Actual disposal of excess facilities can be difficult even for private-sector organizations. Obstacles to disposal include a lack of resources for the upfront planning or the investment necessary to sell or demolish excess facilities, organizational culture, the desire to retain space "insurance" at a local location (sometimes used for storage of underused equipment), or a belief that what is excess capacity today may be needed in the future.

Representatives of IBM and GM emphasized that their organizations were unable to dispose of excess facilities effectively until the effort was managed by a central organization charged with that responsibility. The central corporate-level organization was responsible for identifying excess facilities, identifying the best method for disposition (such as sale or demolition) and preparing an implementation plan. The plan was reviewed with local management before obtaining corporate-level approval for disposal or demolition and approval of the funding needed to execute the plan.[1]

As noted in Chapter 1, federal agencies own thousands of excess and underutilized facilities and this poses a risk of fiscal exposure to the federal government as a whole. To date, the Base Realignment and Closure (BRAC) process that began in 1988 and continues today has been the most far-reaching and ambitious effort to address this issue.

In June 2010, a presidential memorandum titled *Disposing of Unneeded Federal Real Estate* was issued to address excess properties in civilian agencies.[2] The memorandum states the following:

> For decades, the Federal Government, the largest property owner and energy user in the United States, has managed more real estate than necessary to effectively support its programs and missions. Both taxpayer dollars and energy resources are being wasted to maintain these excess assets. In addition, many of the properties necessary for the Government's work are not operated efficiently, resulting in wasted funds and excessive greenhouse gas pollution. . . . Past attempts at reducing the Federal Government's civilian real property assets produced small savings and had a minor impact on the condition and performance of mission-critical properties. These efforts were not sufficiently comprehensive in disposing of excess real estate and did not emphasize making more efficient use of existing assets.

That presidential memorandum states that federal agency actions, as permitted by law, should include reducing cycle times for identifying excess assets and disposing of them; eliminating lease arrangements that are not cost effective; pursuing consolidation opportunities within and among agencies in common asset types (such as data centers, office space, warehouses, and laboratories); increasing occupancy rates in current facilities through innovative approaches to space management and alternative workplace arrangements, such as telework; and identifying offsetting reductions in inventory when new space is acquired. Federal agencies are also directed to take immediate steps to make better use of remaining real property assets as measured by utilization and occupancy rates, annual operating cost, energy efficiency, and sustainability.

Those actions are intended to result in at least $3 billion in cost savings by the end of FY 2012. An additional $9.8 billion in savings is expected to be realized through the Department of Defense's BRAC efforts from FY 2010 to FY 2012,

---

[1]GM has several examples in which the salvage revenue from demolition exceeded the cost of demolition. In at least one instance, the salvage cost of the structural steel alone exceeded the total cost of demolition.

[2]The full text of the memorandum is available at http://www.whitehouse.gov/the-press-office/presidential-memorandum-disposing-unneeded-federal-real-estate.

of which $5 billion is a direct result of reducing operating and maintenance costs by disposing of excess facilities or other consolidation efforts.

Three previous National Research Council reports have addressed various aspects of the disposition of federal facilities. *Stewardship of Federal Facilities: A Proactive Strategy for Managing the Nation's Public Assets* made the following recommendation (NRC, 1998, p. 7):

> Long-term requirements for maintenance and repair expenditures should be managed by reducing the size of the federal facilities portfolio. New construction should be limited, existing buildings should be adapted to new uses, and the ownership of unneeded buildings should be transferred to other public or private organizations. Facilities that are functionally obsolete, are not needed to support an agency's mission, are not historically significant, and are not suitable for transfer or adaptive reuse should be demolished whenever it is cost effective to do so.

Investments in Federal Facilities: Asset Management Strategies for the 21st Century made two additional recommendations on this topic as follows (NRC, 2004a, p. 5):

> Recommendation 2(c). To facilitate the alignment of each department's and agency's existing facilities portfolios with its missions, Congress and the administration should jointly lead an effort to consolidate and streamline government-wide policies, regulations, and processes related to facilities disposal, which would encourage routine disposal of excess facilities in a timely manner.

> Recommendation 2(d). For departments and agencies with many more facilities than are needed for their mission . . . Congress and the administration should jointly consider implementing extraordinary measures like the process used for military base realignment and closure (BRAC), modified as required to reflect actual experience with BRAC.

A third study, *Intelligent Sustainment and Renewal of Department of Energy Facilities and Infrastructure* (NRC, 2004b), outlined a decision-making process that could be used by the Department of Energy to determine whether to repair, renovate, or replace facilities and to determine whether a facility should be retained or disposed of (Figure 4.1). The process outlined may be of use to other federal agencies.

If the disposal of excess and underutilized properties by civilian agencies can be successfully implemented, the federal government would reduce its risk of fiscal exposure related to the ownership of buildings, reduce its total operating costs, and reduce its long-term maintenance and repair requirements and costs. Disposal of excess and underutilized facilities would also help to meet other public policy objectives related to reductions in the use of energy and water and in greenhouse gas emissions. To realize those long-term savings and benefits, a coordinated, sustained, and funded effort will be required.

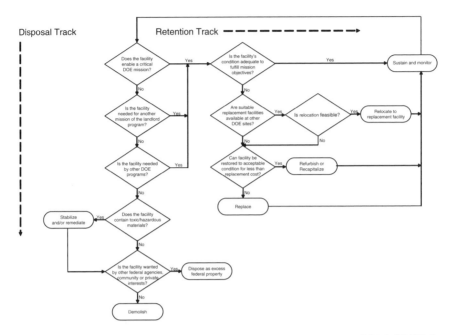

FIGURE 4.1 Process for decision-making to repair, renovate, or replace. SOURCE: NRC, 2004b.

PURSUE A PROACTIVE STRATEGY TO MINIMIZE THE TOTAL FACILITIES "FOOTPRINT"

In addition to disposing of excess and underutilized facilities, private-sector and other high-performance organizations proactively initiate strategies to minimize their total facilities footprint and the associated costs (such as those for equipment, furniture, and landscaping). As long as a facility is occupied or otherwise in use, even if it is underutilized, electrical, mechanical, life-safety, and other systems must be kept in safe operating condition. Providing services to unneeded space is not cost-effective.

As noted in Chapter 1, advances in technology are changing the concept of workplace. Alternative work strategies, such as telecommuting, offer the potential for both public and private-sector organizations to reduce their required amounts of office or administrative space substantially. Doing so can also reduce their overall maintenance and repair requirements.

IBM was described as a company that had nearly gone out of business in 1992 and had chosen to rebuild itself. Recognizing that the nature of work had changed over the past 5 years, IBM has instituted a strategy to reduce its facilities requirements. Four out of five IBM consultants now work at client sites, at home, or only occasionally at IBM facilities, primarily for group meetings. Almost half its employees involved in support functions (for example, human resources, procurement, administration, or finance) work primarily from home. Because IBM depends on rapid communication among its staff and to ensure continued productivity, the company has invested heavily in monitoring and surveillance technologies to track the work that is being performed off-site. Performance-based contracts between IBM and its employees drive variable pay components each year (St. Thomas, 2010).

As leases expire, IBM reduces its total amount of leased space. It is also changing the type of space that it leases from dedicated offices to meeting spaces, team rooms, and conference rooms. In one of its locations, that strategy resulted in a 50 percent reduction in total leased space.

Some federal agencies, such as the U.S. Patent and Trademark Office (USPTO) have been using alternative work arrangements for more than 5 years. The USPTO has been able to measure the results in productivity (for example, sick days taken and number of patent applications examined) and employee retention (USPTO, 2010; Campbell, 2011). In a recent presentation to the Federal Facilities Council, representatives of the USPTO reported that the agency has been able to reduce its total amount of leased space and has avoided leasing costs of almost $20 million (Campbell, 2011). Other agencies, such as the General Services Administration, are implementing alternative work arrangements to reduce their demand for office space and to reduce operating, energy, water, maintenance, and repair costs.

The Telework Enhancement Act of 2010 (PL 111-292) grants federal employees eligibility to telework and requires all federal agencies to establish telework policies. As these policies are implemented, there will be opportunities to reduce the federal facilities footprint further. If successful, such strategies could result in substantial reductions in long-term maintenance and repair requirements and a more sustainable portfolio of federal facilities.

CORRELATE THE EFFECTS OF FAILURE WITH THE ORGANIZATION'S MISSION

The primary objective of portfolio-based facilities management is to ensure that facilities-related investments enable the organization's mission. Private-sector companies, such as GM and IBM, which produce vehicles and computers, respectively, have been able to correlate the failure to invest in maintenance and repair with their organizations' mission, which is to make a profit for their owners and shareholders.

In the automotive industry, profit is realized by producing and selling vehicles at a sales volume that minimizes such overhead costs as engineering and design and increases profit per unit. At a manufacturing plant, overhead per vehicle is reduced by maintaining the design throughput of the plant, which is about 60 to

75 vehicles per hour, or one vehicle per minute or better. At that level of production, an 8-minute system failure (downtime) caused by lack of maintenance of electrical systems or equipment can reduce a plant's output by 8 to 10 vehicles, which can be easily quantified in terms of lost sales and in turn, lost profit. For the facility manager, the goal is to ensure that the facility components and equipment that are critical to production are kept in such a condition as to ensure reliable performance. Similarly, at IBM, a relatively short failure (outage) on the production line caused by unreliable equipment or infrastructure systems can result in production of fewer computers and cost the company millions of dollars in lost sales.

When a failure does occur, companies like GM conduct a root-cause analysis to determine why it happened, determine the appropriate solution, and share the lesson learned with other plants that may be at risk from similar failures. For a component failure that puts the organization's mission at risk, the issue and the solution are rolled up to the corporate level, where a budget line item is submitted to fix or replace the component across the organization. The credibility of such requests is supported by evidence generated through the root-cause analysis.

Root-cause analysis is a key component of reliability-centered maintenance (RCM), the real-time monitoring of the performance and condition of facility systems and equipment. RCM has been used extensively in the aircraft, space, defense, and nuclear industries, in which functional failures can result in loss of life, can have national security implications, or can have extreme environmental effects (NASA, 1996; NRC, 1998). A rigorous analysis of failure modes and effects is used to determine the appropriate type and timing of maintenance of systems and equipment (for example, preventive maintenance, predictive testing and inspection) (NRC, 1998). The overall objectives are to ensure that the performance and service lives of systems, equipment, and components are optimized and to ensure that critical elements are replaced before they fail because of wear and tear.

Because the products and programs of federal agencies are typically related to protecting the public's health, safety, and welfare, not to making a profit, the links between the failure of a system or component, maintenance and repair investments, and organizational missions are more difficult to convey to decision-makers and others. Nonetheless, several agencies have developed or are developing approaches for doing so. Those approaches include the use of RCM, the development of a mission dependency index (MDI), and the development of a risk-based process for maintenance and repair investment decision support. They are described below.

• **Reliability-Centered Maintenance (RCM).** The National Aeronautics and Space Administration (NASA) began applying a version of RCM to some of its facilities systems and equipment in the 1990s. In NASA, the RCM approach integrates reactive maintenance (run-to-failure or breakdown maintenance), preventive (interval-based) maintenance, predictive testing and inspection (condition-based) and proactive maintenance. Those four maintenance strategies are applied to systems and equipment on the basis of the consequences of equipment failure and the potential effects on mission, safety, environment, and life-cycle cost (NRC, 1998).

The Smithsonian Institution has more recently implemented a version of RCM. The Smithsonian's RCM approach is to perform the right maintenance on the right equipment at the right time. "Right" entails a balance of resources and maintenance techniques developed from industry best practices. To institute its RCM program, the Smithsonian hired an expert consultant and set out immediately to deploy the most relevant RCM technologies to conduct condition-based tasks. An analysis looked at the following:

- The functions and performance standards of the equipment that will be maintained,
- The failure modes of that equipment,
- The causes of each failure,
- The effects of failure,
- The tasks that could predict or prevent failure, and
- Workarounds if there are no proactive tasks to mitigate failure.

Templates were developed for systems and equipment with similar failure modes. Decisions about the type of maintenance to be performed on specific types of equipment were based in part on whether designed bypasses or other redundancies were built into the system and whether spare parts were on hand or readily available and economical to purchase.

A number of nondestructive technologies are used for predictive testing and inspection (PT&I) of equipment. PT&I minimizes downtime by allowing the Smithsonian staff to conduct investigations when the systems are operating. Data are analyzed to identify measurements that exceed known threshold values and to identify changes in condition. The analysis is used to plan and schedule repair or replacement of equipment or components before they fail. All PT&I data and findings are entered into the Smithsonian's computerized maintenance management system to build an equipment and system history, which helps further to detect problems before they become serious.

- **Mission-Dependency Index.** The mission-dependency index (MDI) was developed jointly by the U.S. Navy, the U.S. Coast Guard, and NASA as a process for incorporating operational risk management into facilities asset management (Antelman et al., 2008). The MDI is a severity metric (on a scale of 1 to 100) for risk that considers the ability to relocate a mission to another facility and the ability to withstand mission interruption. That is, if a facility or component is deemed not usable for mission accomplishment, how long will the mission be interrupted (minutes or days?) and can the mission be moved elsewhere (Is it impossible or easy?). The MDI model considers both facility intradependency (facilities are controlled by the mission stakeholder) and facility interdependency (facilities are needed, but not controlled by the mission stakeholder). To develop an index, an agency must first survey the various stakeholders to collect the relocation and interruption data that are specific to the agency. Through weights established for both data elements, an algorithm derived from a regression analysis is used to compute the MDI. The index identifies which facilities or components are mission-critical, mission-supportive, or otherwise categorized. Once the analysis is completed for an agency, the MDI can be used to help set priorities among proposed maintenance and repair activities and to assess vulnerability, which could be a mission loss predictor.

• **U.S. Army Corps of Engineers.** The U.S. Army Corps of Engineers (USACE) is developing a portfolio-based, risk-related approach for maintenance and repair investments in its inland navigation program and other civil-works programs. The approach is an integrative one that considers all facilities-related investments, including capital, operations, and maintenance and repair.

The goal of the inland navigation program is to provide safe, reliable, efficient, effective, and environmentally sustainable waterborne transportation systems for commerce, national security needs, and recreation. The infrastructure and components required to achieve those objectives include locks, dams, channels, and canals, gates, valves, operating mechanisms, controls, and machinery. Failure of one of those components due to lack of maintenance and repair can result in substantial disruption of traffic and commerce on major waterways, such as the Mississippi River, which can result in economic losses to private-sector companies and communities.

The USACE has established a set of investment objectives and performance measures for the navigation program that are related to the civil-works strategic plan (Table 4.1). It has also developed a set of budget strategies and ranking criteria that are designed to link investments to the mission of the inland navigation program (Table 4.2) and established consequence categories and consequence-rating criteria (Table 4.3).

CORRELATE REPAIR DELAY WITH SUSTAINMENT COST

As noted in Chapter 2 and shown in Figure 2.2, timely investment in maintenance and repair can ensure that facilities systems and components operate efficiently throughout their service lives. Conversely, delaying repairs of facilities can shorten service life and result in an increase in sustainment cost, which is defined as the sum of maintenance cost and renewal cost. The service-life models and the IMPACT model described in Chapter 3 can be used to quantify the costs of delaying repairs.

For agencies that are not using such models, it may still be possible to estimate the costs of delaying repairs with an approach described by representatives of GM. GM centralized the facility-support team for specific groups of facility components (Table 4.4) so that the expert on a given component could build the financial arguments needed to identify the cost of delaying repairs.

TABLE 4.1 Navigation Objectives and Performance Measures

Program Objectives	Performance Measures
Invest in navigation infrastructure when the benefits exceed the costs	Benefit-cost ratio (a project specific measure) Annual net benefits
Enhance life-cycle infrastructure management. Improve the reliability of water resources infrastructure by using a risk-informed asset management strategy	Percentage of navigation asset inventory with recent structural or operational risk assessments, including SPRA assessments
	Percentage of navigation asset inventory risk assessments that reveal a significant level of risk
	Number of funded actions under way that address assets regarding which there is a significant level of risk
Operate and manage the navigation infrastructure to maintain justified levels of service in terms of the availability of high-use navigation infrastructure (waterways, harbors, and channels) to commercial traffic	Risk and reliability: facility condition assessment and efforts

SOURCE: USACE, 2010.

TABLE 4.2 Navigation Budget Performance Measures

Budget Strategy	Ranking Criteria
Initiate and complete replacements and rehabilitations	Inland Waterways Users Board priority
	Relative risk of failure
Operations—ensure that projects perform as designed (O&M)	Cumulative benefits
	Cumulative O&M costs for above benefits (over set period)
Maintenance—ensure that projects are safe to operate (managing risk) (O&M)	Navigation channel availability
	Lock closures exceeding 24 hours and one week in duration because of mechanical failure— scheduled and unscheduled
	Condition assessment and consequences or impact
	Cumulative benefits
	Cumulative O&M costs for above benefits (over set period)

SOURCE: USACE, 2010.

TABLE 4.3 Inland Navigation Consequence Categories and Consequence Rating Criteria

Consequence Category	Consequence Rating Criteria
1	Maximum risk to mission Highest economic loss: more than 5 billion ton-miles Probable life-safety effect Minimum acceptable operations service level Court-mandated action Shutdown of energy generation or distribution facilities (such as power plants and oil distribution facilities) for national public use with no alternative modes of transportation
2	High risk to mission No life-safety effect High economic loss: 2.5 billion to 5 billion ton-miles Diminished cost efficiency of energy generation or distribution facilities (such as power plants and oil distribution facilities) for national public use with higher-cost alternative modes of transportation
3	Moderate risk to mission No life-safety effect Moderate economic loss: 1 billion to 2.5 billion ton-miles
4	Low risk to mission No life-safety effect Low economic loss: 500 million to 1 billion ton-miles
5	Negligible risk to mission No life-safety effect Least economic loss: less than 500 million ton-miles

SOURCE: USACE, 2010.

TABLE 4.4 Categories of Facilities Components Used by General Motors

CORPORATE PROGRAMS	GENERAL FUNDS
Roofing	General lighting
Paving	Parking lot lighting
HVAC	Sewers
Fire protection	Water supply
Fork truck batteries	Piping distribution
Electrical substations	Floors
Wastewater treatment plants	Sash and glass
Truck docks and doors	Elevators
Powerhouse, including compressed air, chillers, and cooling	Café
Computer rooms	Restrooms
Scrap conveyors and cranes	Plant administration buildings
Railroad tracks	Underground and aboveground storage tanks
Externally driven environmental and regulations	Signs
Decommissioning or legacy costs	Landscaping

SOURCE: McNabb, 2010.

For example, the cost to repair or replace heating, ventilation, and air-conditioning (HVAC) units is X. If the decision to replace or overhaul them is deferred for another year, the cost will be X + I + O where I is the added cost created by inflation for 1 year converted to net present value and O is the additional operational cost incurred because the equipment was not operating as efficiently as it was designed to do (for example, additional energy was consumed because of improperly firing burners, bad dampers, ineffective controls, or excessive pressure drop on filter banks). If I + O exceeds X, repairs should proceed immediately to avoid an increase in sustainment cost. When presenting the results of such an analysis to senior decision-makers, facilities program managers also identify other outcomes from the proposed investment, such as employee comfort or improved safety.

REMOVE MUST-FUND PROJECTS AND THOSE SUPPORTED BY ACCEPTABLE FINANCIAL PAYBACK FROM THE MAINTENANCE AND REPAIR ACCOUNT

Each of the private-sector organizations interviewed for this study identified projects that must be funded to manage risk to the organization and funded them through sources other than the maintenance and repair account. They also seek out opportunities to use "must-fund" projects to realize long-term operating efficiencies.

For instance, projects that are required for an organization to be in compliance with government regulations (such as worker health and safety and air and water-quality regulations) are considered must-fund projects. That is because the risks associated with noncompliance—large fines or legal action—are much greater than the costs of the projects themselves. Similarly, activities to fix conditions that present a hazard to workers or others are viewed as must-fund projects to manage the risks associated with insurance claims and lawsuits. Must-fund projects are taken out of the maintenance and repair account and funded from other discretionary sources, such as the operating account.

At the same time that must-fund projects are identified, facilities managers look for opportunities to support operational efficiencies and to reduce long-term costs. For example, if the regulatory standards for the treatment of effluent water are no longer achievable with existing, worn-out or technologically obsolete equipment, there may be an opportunity to replace it with more technologically advanced equipment. Even if the organization does not have to comply with new regulations immediately, it frequently makes good business sense to replace the equipment sooner rather than later with equipment that will allow the organization to meet the expected regulations instead of spending additional funds to maintain and repair obsolete equipment.

Private-sector firms also indicated that maintenance and repair projects that have a short payback time (such as 1 to 3 years) are funded on their own merit from sources other than the maintenance and repair account. For example, energy-saving projects, such as relamping lighting fixtures with more efficient components, can quickly pay back the original investment and continue to generate benefits for additional years. Because such projects are justifiable on their merit, they do not compete with other projects for maintenance and repair funds.

In the federal government, maintenance and repair funds for most agencies are part of the general operations account and are not earmarked specifically for maintenance and repair projects (NRC, 1998). "Structuring the account in this way accommodates overlaps between work, operations and alterations. For example, equipment operators often do routine equipment maintenance and alteration projects, including work that could be considered repairs" (NRC, 1998, p. 28). Federal agencies could also identify must-fund projects and fund them from the operations side of the account rather than from the maintenance and repair side.

Public-private partnerships (PPPs) can also be used to secure third-party financing to accomplish some types of maintenance and repair projects or activities. PPPs are contractual agreements between public-sector and private-sector organizations wherein the private-sector organization, in exchange for compensation, agrees to deliver services or even facilities that could be provided by a public-sector organization (Keston Institute, 2011). Through the use of PPPs, federal agencies can implement necessary capital or maintenance and repair requirements through third-party financing and can gain access to private-sector expertise. PPPs, however, are not without risks and the risks need to be accounted for.

Energy-savings performance contracts (ESPCs) are one type of PPP that many federal agencies are already using to leverage available funding. Under such contracts, an energy service company (ESCO) typically conducts a comprehensive energy audit for groups or types of facilities and identifies improvements that could save energy. In consultation with the owner agency, the ESCO designs and constructs a project that meets the owner's needs and arranges the necessary financing. The ESCO guarantees that the improvements will generate energy-cost savings sufficient to pay for the project over the term of the contract. After the contract ends, all additional cost savings accrue to the owner organization.[3] The key feature of this model is that the private sector provides front-end funding for the project in return for the ability to receive benefits from future savings. In this way, the risk associated with nonperformance is shifted to the private-sector partner. The general concept is similar to the paid-from-savings approach promoted by the U.S. Green Building Council.[4]

About $2.3 billion has been invested in federal facilities through ESPCs. The ESPC projects contain guarantees that will result in $6 billion in avoided energy

---

[3]See Federal Energy Management Program Web site http://www1.eere.energy.gov/femp/financing/espcs.html.

[4]Additional information on this approach is available at http://www.usgbc.org/DisplayPage.aspx?CMSPageID=2204.

costs over the life of the contracts. The contracts have also resulted in savings of 18 trillion British thermal units—roughly equivalent to the energy used by a city of 500,000 people (Kidd, 2010).[5]

Performance-based maintenance contracts (PBMCs) are another type of public-private partnership. Like ESPCs, PBMCs engage private-sector firms to provide long-term maintenance, repair, and replacement work.

When contractors perform maintenance and repair activities for federal agencies, the contracts typically specify the procedures and materials to be used. As long as the contractor meets the specifications, the risks associated with the contract are fully retained by the agency. In contrast, in a PMBC, the agency specifies performance goals. The contractor is free to select the methods, materials, and timing of maintenance actions to meet those goals, but also assumes the risks associated with failure to meet those goals.

The benefits of PBMCs include the ability to obtain financing from the private sector for expensive repair and reconstruction work, the flexibility with which contractors can exploit advances in methods and materials without the need to renegotiate contract terms, and the potential for transferring knowledge of innovative practices from the contractors to the agencies. However, to achieve those benefits, federal agencies must clearly and carefully identify the performance goals that are to be met, provide appropriate incentives so that contractors will take appropriate measures before systems or components fail, and regularly monitor implementation of the contract. Failing to do those things can result in a failure to realize the expected benefits. Agencies also take on a long-term liability (performance payments) that can become rather large in the case of a portfolio of facilities and can create the risk of fiscal exposure (TRB, 2010).

Other types of public-private partnerships potentially could be used to lower the costs and risks associated with facility ownership. For example, in a design-build-finance-operate contract the private-sector partner finances, designs, constructs, and operates a facility under a long-term lease and the public organization takes ownership of the property at the end of the lease. The private-sector partner assumes the financial risks related to project delivery, maintenance, and revenue; and the public-sector partner assumes all the risks related to facilities ownership, once it takes over the facility.

A private finance initiative (PFI) is a PPP-based arrangement used in the United Kingdom. In PFI contracts, the private-sector partner provides funding and delivers the public facilities and infrastructure based on the "output" (performance) specifications. The public sector does not own the facility, but reimburses the private-sector partner with a stream of committed payments for the use of the facility over the period specified in the contract (Allen, 2001). However, the payments are conditional on the ability of the private-sector partner to meet the per-

---

[5]Additional information on ESPCs is available at http://www1.eere.energy.gov/femp/financing/espcs.html.

formance specifications (standards); the specifications address the strategic needs of the facility owner and occupants.

Chapter 5 of the report *Investments in Federal Facilities: Asset Management Strategies for the 21st Century* (NRC, 2004a), describes and evaluates various funding approaches for acquiring federal facilities. It offers the following recommendation (NRC, 2004a, p. 10):

> Recommendation 11: In order to leverage funding, Congress and the administration should encourage and allow more widespread use of alternative approaches for acquiring facilities, such as public-private partnerships and capital acquisition funds.

STRATEGICALLY ASSESS THE CONDITION OF FACILITIES

Private-sector companies and government organizations conduct or contract for some facility condition assessments to provide the information necessary to develop a credible maintenance and repair request. As noted in Chapter 3, technology can also be used to monitor the condition of systems and equipment in real time.

To realize other operational efficiencies in conducting condition assessments, such organizations as Marriott and GM perform most of their facility assessment at a corporate level according to type of component (such as roofs, HVAC systems, and electricity distribution). For example, an expert in roofing-condition analysis will visit all plants in a region within a short period to minimize travel expenses. The assessment may be spread over a 4-year cycle so that about 25 percent of the facilities are assessed each year. At the completion of assessment, an analysis is conducted to determine whether the average life of roofs is improving or deteriorating. The resulting information is used to set priorities for roof repairs throughout the organization, ensuring that facilities whose roofs are in the worst condition are addressed first. The corporate roofing expert also trains local staff during the assessment visits, advising them of potential changes in roof-maintenance practices, of the latest trends in roofing technology, and of which roofing type is most cost-effective for their climate and particular plant conditions.

The National Nuclear Security Administration (NNSA) has implemented a similar process for the inspection, management, and maintenance of its 16 million square feet of roofs as an element of its facilities and infrastructure revitalization program. In a presentation to the Federal Facilities Council, it was reported that the program has resulted in improved condition of the NNSA's roofing portfolio; in increased average remaining service life; in the replacement of 3 million square feet of roofs with more energy-efficient, sustainable roofs, including 2 million square feet of cool roofs; in $17 million of savings in overhead costs; and in other benefits (Moran, 2010).

CONDUCT A YEAR-END BUDGET REVIEW TO EVALUATE INVESTMENT PERFORMANCE

Investments in Federal Facilities: Asset Management Strategies for the 21st Century stated that (NRC, 2004a, p. 69):

> Continuous evaluation and feedback on processes and investments are essential to controlling and improving them. Feedback can be positive or negative, take many forms, and be used over various timescales. . . . In best-practice organizations, the performance of projects, processes, people, business units, physical assets, investments, and the organization as a whole are continuously monitored and evaluated over both the short and long term using performance measures and a variety of feedback processes.

In addition to a variety of feedback processes described in Chapter 4 of the 2004 report, the conduct of a year-end evaluation of budget performance was identified during the course of the present study. This concept was used by at least one of the facility organizations at GM. The evaluation compared the "submitted budget" the "approved budget" and the "actual funding spent" by line item. Where there was a substantial variance (greater than 10 percent), the root cause was analyzed by using the same basic process as would be used for analyzing an equipment failure. The root cause of the deviation was shared with senior decision-makers throughout the organization. GM facilities managers found that this type of analysis improved their future budget submissions and increased their credibility with other decision-makers including those who approved the budget.

5

Communicating Outcomes and Risk

Obtaining funds to maintain and repair the federal government's buildings and infrastructure has long been challenging (GAO, 2008; DOD, 2001). Senior executives of federal agencies and departments are inundated with requests to support funding for a wide array of mission-related programs, each accompanied with compelling messages and evidence of the need for funding. In 2011 and beyond, the challenge of making a compelling case for investment in facilities maintenance and repair may be greater than at any time in the recent past, given the current fiscal outlook (NAPA and NRC, 2010; GAO, 2011c).

Federal facilities program managers who use an outcomes-based approach for developing maintenance and repair funding requests will need to communicate the outcomes and the basis of their development persuasively to other decision-makers and colleagues. Although carefully designed and implemented communications to staff and upper management will not guarantee success in obtaining the required funding for maintenance and repair activities, poor communications are likely to doom such a request.

ISSUES RELATED TO EFFECTIVE COMMUNICATION

Communication is the science and practice of transmitting information in a manner that succeeds in evoking understanding (NRC, 2004a). Effective communication is more than a good presentation or a dynamic messenger: it is about the quality of the message, the credibility of the information, and the deliberations that ensue (NRC, 2004a).

Between the formation of the present committee in November 2009 and its first meeting in December 2009, three events illustrated the importance of effective communication in evoking understanding of and securing support for a proposal. In one instance, U.S. public-health officials were criticized for not producing sufficient H1N1 flu vaccine and for not insisting on its use (Klass, 2009). In two other instances, proposals to change mammography protocols were criticized (Goodman, 2009; Kolata, 2009), and a program to install smart electric meters in homes was found to upset many consumers (Wald, 2009). Those three examples were not strictly speaking communication failures. However, the failure to communicate clearly made complicated proposals difficult for the public to understand and consequently difficult for the public to accept and support.

In comparison, the maintenance and repair of federal facilities is a less visible issue, unless there is a system failure that leads to a death, serious injuries, substantial economic or property losses or public embarrassment. Nonetheless, effective communication is a necessary ingredient for making a compelling case for funding maintenance and repair activities.

The barriers to effective communication include "lack of a common terminology; lack of trust in the source of information; poor interpersonal relationships; differing individual and group values; and unexpressed assumptions" (NRC, 2004a, p. 63).

Lack of a common terminology can easily lead to miscommunication about the purposes and anticipated outcomes of investments in facilities maintenance and repair. In the federal government, there is a great deal of variance in the terminology related to maintenance and repair activities, in the measures used for outcomes, in the definitions applied, and in the thresholds used to determine what activities fall into a particular budget category. When agencies are communicating with congressional committees, with the Office of Management and Budget, with other oversight groups, and even among themselves, that variance results in inconsistent and conflicting messages, which cause confusion. Confusion, in turn, leads decision-makers to call to question the credibility of the messenger and the message.

Effectively communicating the links between outcomes of maintenance and repair investments and an organizational mission has also proven to be difficult. In part, that is due to the difficulty of predicting rates of failure of facilities systems or components, the difficulty of predicting remaining service life, variation in costs of maintenance and repair of specific systems and their components, and the difficulty of quantifying the adverse consequences of potential failures.

Typically, three predominant approaches are used by federal program managers to calculate required maintenance and repair funding: a percentage of the current replacement value of the entire facilities portfolio; a sustainment model such as that used by the Department of Defense; and the total cost of deferred maintenance and repair projects. Although industry experience and practice are sufficient to support the applicability of those approaches, senior decision-makers may not find them compelling.

For example, it is not intuitively obvious how a request for 2 to 4 percent of the current replacement value of a portfolio of facilities will contribute to meeting an agency's mission. Nor is it obvious how a statement that a department has a backlog of billions of dollars of deferred maintenance and repair projects will motivate a decision-maker. In fact, agency presentations to the committee indicated substantial negative reactions by senior decision-makers to methods based only on deferred maintenance information. A strong negative message may also lead senior decision-makers to believe that investment in maintenance and repair is not worth addressing unless there is a direct health, safety or legal compliance issue (Koren and Klein, 1991; Siegrist and Cvetkovich, 2001). Research also shows that a historical record of volatility and lack of predictability is likely to reduce support for investment (Weber et al., 2005).

A related issue is persuading decision-makers and others of the importance of maintenance and repair investments to prevent actual failures of systems or components. Despite difficult conditions, federal facilities personnel do their best to keep deteriorating systems running through work-arounds. The result is that systems seldom fail in a highly visible manner, so the risks associated with deteriorating systems and the benefits of timely investments in maintenance and repair are not readily apparent to decision-makers and the urgency of investment can be difficult to convey.

COMMUNICATION STRATEGIES USED BY
PRIVATE-SECTOR ORGANIZATIONS

As noted in Chapter 4, the private-sector organizations whose representatives were interviewed for the present report were able to secure adequate funding for maintenance and repair for a variety of reasons. In presentations to the committee, the organizations were explicit about the importance of effective communication for receiving that support.

Each private-sector facilities management organization presented a clear message on the management of maintenance and repair requirements and budget. That has allowed them to develop the practices and the understanding—the culture—of what such investment means to a company as a whole and an understanding of the processes and procedures involved. This translates into a consistent message that can be understood by decision-makers at all levels of the organization. All of these practices are consistent with other best-practice organizations, which do the following (NRC, 2004a, p. 2):

> Establish a framework of procedures, required information, and valuation criteria that aligns the goals, objectives, and values of their individual decision-making and operating groups to achieve the organization's overall mission; create an effective decision-making environment; and provide a basis for measuring and improving the outcomes of facilities investments. The components of the framework are understood and used by all leadership and management levels.

The framework of procedures helps to align the values and objectives of different groups in the organization. Components of the framework include common terminology, a business case analysis, and clearly defined evaluation processes that incorporate multiple decision points (NRC, 2004a).

Two clear messages that came out of the private-sector presentations dealt with the need to address two points of context and content: total cost of ownership and adding to the competitive edge.

When communicating about the total cost of ownership, facilities program managers in private-sector organizations discussed maintenance and repair activities in terms of system and component renewal, sustainment, planning, disposal, and life-cycle costing. In doing so, they implied that maintenance and repair investments and activities should be well integrated with current operations. As noted in Chapter 4, one way to do that was to identify safety and potential regulatory violations, vet them, screen them, and present them to senior decision-makers as must-fund requirements that should be paid for out of operations accounts.

The Association of Higher Education Facilities Officers-APPA, which focuses on facilities owned by colleges and universities, has developed a framework for integrated decision-making that also looks at managing the total cost of ownership of a facilities portfolio. Total cost of ownership is divided into three separate categories: nonrecurring costs (such as costs of planning and construction of new buildings and additions); annual recurring costs (such as costs of maintenance, operations, repairs, and utilities); and periodic recurring costs (such as costs of recapitalization, remodeling, and replacements) (Rose et al., 2007). The framework was developed, in part, to make it clear to university presidents and other decision-makers how investment in maintenance and repair affects the overall value of the facilities portfolio.

With regard to adding to the competitive edge and the bottom-line profit message, facilities program managers in private-sector organizations spoke about the relationship of facilities to workforce recruitment, risks to missions, and the alignment of facilities to operations—referred to as right tasks plus right skills plus right places. That approach is consistent with portfolio-based facilities management, which treats facilities as enablers of missions that can contribute to an organization's competitive edge, as opposed to being simply a cost of doing business. Effective communication and links to organizational objectives were demonstrated by the presentation of allocation models that began with deliberative assessments of current and future needs and flowed to funding requirements and company benefits. Various feedback loops tied organizational levels together and provided multiple decision points and opportunities for reevaluation and adjustments of strategies as necessary.

As noted in Chapter 4, one way that some private-sector organizations link maintenance and repair requirements to the bottom line is to group maintenance and repair projects by component, such as roofs, heating ventilation, air-conditioning, and fire protection. Grouping all the projects related to a component type makes the benefits and risks associated with projects more transparent to decision makers. It also allows them to more easily vet, screen, treat, and then set priorities among the projects that should be funded first to ensure that the organization's profits are not adversely affected by unreliable systems or components.

The private-sector organizations also touched briefly on the importance of building relationships and trust within an organization. Trust—unquestioning belief in and reliance on someone or something—is important for the success of almost all forms of human interaction (NRC, 2004a). Trust is built among decision making and operating groups in organizations by ensuring that everyone has access to the same information. A 2004 National Research Council study found that "facilities management operating groups had gained or retained credibility and built trust at the institutional level by providing sound information, by incorporating rigor into their analyses, by giving high-quality presentations, and by submitting realistic, reasonable requests for investment proposals" (NRC, 2004a, p. 64).

That federal agencies do not have a single integrative bottom-line focus complicates their efforts to build a strong message. Nevertheless, theory and practice suggest that the value of investing in maintenance and repair activities can be more effectively communicated than it has typically been.

COMMUNICATING THE VALUE OF MAINTENANCE AND REPAIR TO A MISSION

Communication theory emphasizes three elements of persuasive communication: identifying shared objectives, defining the approach and acknowledging others, and supporting the approach with solid research (Bettinghaus and Cody, 1994).

Federal facilities program managers who seek greater funding support for maintenance and repair activities should show senior decision-makers precisely how a proposed request for funding meets the objectives of the entire organization, not just the objectives of the facilities management group. Given the reality that senior decision-makers often stay in their positions for only a few years, it is prudent to present the results as outcomes that are directly tied to explicit and implicit missions and other public policy objectives for which senior decision-makers will be held accountable.

To gain more support for maintenance and repair investments, federal facilities program managers will also need to communicate that there is a disciplined and deliberate approach for funding requests, that requests will result in outcomes that are directly tied to their organization's mission, and that the funds received will be invested effectively to achieve the predicted outcomes. Federal facilities program managers will also need to track how the funds are invested and report the resulting outcomes in comparison with the predicted outcomes. The approach embedded in the Mission Dependency Index and the approach used by the U.S. Army Corps of Engineers described in Chapter 4 illustrate how maintenance and repair requests can be clearly and effectively tied to a mission.

Senior decision-makers will also want to know the return on investment or expected value of investment. Facilities program managers should understand and be able to communicate effectively the economic value of a component or system to a mission, and the cost of protecting its value. To do that they will need to identify the types of deterioration or other adverse events that will lead to loss of mission, the vulnerabilities of facilities to the adverse events, the potential loss of economic value if a failure occurs, the accumulation of potential losses until the system is repaired, and how vulnerabilities can cascade into additional failures. For example, facilities program managers should be able to identify the consequences if a component in a heating system causes the entire system to go down for 2 days in the middle of winter or if a roof leaks or collapses and interrupts research or other activities or destroys computer equipment. They also need to identify what can be done to prevent such situations, how much it will cost to avoid the risk, and how much it will save in other costs.

Program managers will need to be able to characterize and explain the level and types of uncertainty inherent in a funding request. Uncertainty is the lack of sufficient information to describe an existing situation (such as unpredictability of a budget) or projection (such as remaining service life). They will need to communicate their level of confidence in the information that they are presenting and identify unavailable information that if available could affect the prediction of outcomes.

The literature offers several suggestions for increasing the chances of successful communication about maintenance and repair investments. A realistic request is the first. Federal facilities program managers should make sure that their outcome estimates can withstand the scrutiny of colleagues and outside experts. Second, transparency is essential: the basis of outcome estimates needs to be available. Senior decision-makers may support an outcomes-based approach, but they will be concerned that the outcomes will not materialize as predicted. Hence, it is prudent for facilities program managers to propose a midcourse evaluation of the outcome measures. They should also be prepared to acknowledge the strengths and weaknesses of this approach in comparison with other options.

Federal facilities program managers should be prepared to explain the value of an outcomes approach because of its complexity. They should expect resistance from some managers on the grounds that an outcomes-oriented set of measures obfuscates a request in an unwieldy sea of numbers. For that reason, they should plan for multiple internal and external communications. No one should expect that a single written deliverable to senior decision-makers will suffice. The challenge is to produce a set of measures of outcomes that will satisfy colleagues and yet be defensible in front of multiple skeptical audiences (Muto et al., 1997). One way for federal facilities program managers to develop consistent messages quickly would be to share lessons learned among agencies about the messages and measures that gained the greatest acceptance with decision-makers and about the messages and measures that created skepticism.

6

Findings and Recommendations

The committee's task was to develop methods, strategies, and procedures to predict outcomes anticipated from investments in federal facilities maintenance and repair. To fulfill its task, the committee was asked to address the following questions:

- Are there ways to predict or quantify the outcomes that can be expected from a given level of investment in maintenance and repair of federal facilities or facilities' systems?
- What risks do deteriorating facilities, deteriorating building systems (such as mechanical and electrical), or deteriorating components (such as roofs and foundations) pose to the achievement of a federal agency's mission or to other organizational outcomes (for example, physical security, operating costs, worker recruitment and retention, and healthcare costs)?
- Do such risks vary by facility type (such as offices, hospitals, industrial, and laboratories), by system, or by function (such as research and administrative)? Can the risks be quantified?
- What strategies, measures, and data should be in place to determine the outcomes of facilities maintenance and repair investments? How can those strategies, measures, and data be used to improve the outcomes of investments?
- Are there effective communication strategies that federal facilities program managers can use to inform decision-makers better about the cost-effectiveness of levels of investment in facilities' maintenance and repair?

Chapters 1 through 5 provide context and address the various aspects of the statement of task. The present chapter extracts the key findings of the report and presents seven recommendations for improving the outcomes of investments in maintenance and repair of federal facilities. Chapter 7 is intended to show how some of the recommendations could be implemented by federal facilities program managers.

FINDINGS

Finding 1. An array of beneficial outcomes can be achieved through timely investments in facilities maintenance and repair (Table 6.1). Those outcomes support mission achievement, compliance with regulations, improved condition, efficient operations, and stakeholder-driven initiatives. All the outcomes can be measured. Some outcomes including reliability and physical condition can be predicted; that is, they can be estimated before an investment is made or if an investment is not made.

When federal facilities program managers identify maintenance and repair requirements, they typically include projects that focus on objectives related to a mission, to compliance with safety and health regulations, to improving facility condition or extending service life, to efficient operations, or to stakeholder-driven

TABLE 6.1 Beneficial Outcomes Related to Investments in Maintenance and Repair

Mission-Related Outcomes	Compliance-Related Outcomes	Condition-Related Outcomes	Efficient Operations	Stakeholder-Driven Outcomes
Improved reliability	Fewer accidents and injuries	Improved condition	Less reactive, unplanned maintenance and repair	Customer satisfaction
Improved productivity	Fewer building-related illnesses	Reduced backlog of deferred maintenance and repairs	Lower operating costs	Improved public image
Functionality	Fewer insurance claims, lawsuits, and regulatory violations		Lower life-cycle costs	
Efficient space utilization			Cost avoidance	
			Reduced energy use	
			Reduced water use	
			Reduced greenhouse gas emissions	

requests. A wide array of beneficial outcomes can result from investments in maintenance and repair, many of which are interrelated. All the outcomes can be measured (as described in Chapter 7), although most of the measures are based on data that are gathered after the fact ("lagging measures"). Some outcomes, including reliability and physical condition (as opposed to financial condition) can be predicted by using processes such as reliability-centered maintenance or by using physical-condition models and indexes that calculate service life and remaining service life. No agency should expect to track all the outcomes. Instead, each agency will need to set priorities among the objectives and outcomes to be achieved through investments in maintenance and repair.

Finding 2. Deteriorating facilities and systems pose risks to the federal government, its agencies, its workforce, and the public. Among them are risks to the achievement of federal agencies' missions; risks to safe, healthy, and secure workplaces; risks to the government's fiscal soundness; risks to efficient and cost-effective operations; and risks to achieving public policy objectives.

Just as investments in maintenance and repair can have an array of beneficial outcomes, the lack of investment and the deferral of needed maintenance and repair projects can pose risks—measure of the probability and severity of adverse events. Adverse outcomes include interruptions in or stoppages of agency operations; accidents, injuries, and illnesses; lawsuits and insurance claims; increased operating costs; shortened service lives of equipment and components; diversion of constrained resources to excess, obsolete, and underutilized facilities; failure to meet public policy objectives; and damage to the federal government's public image.

Risks will increase as federal facilities age and as building systems and components deteriorate through wear and tear and lack of investment in maintenance and repair. (Risks can also be related to geography, climate, and other factors.) Risks associated with deteriorating facilities and systems include risks to federal agencies' missions; to safe, healthy, and secure workplaces; to the government's fiscal soundness and public image; to efficient operations; and to the achievement of public policy objectives related to energy independence and environmental sustainability.

Finding 3. The risks associated with deteriorating facilities vary by type of facility, by system, by existing condition, by function, by utilization, and, most important, by the relationship of the facilities to an agency's mission. Risks can be identified qualitatively and some can be quantified.

The missions and programs managed by federal departments and agencies vary widely. Such variation means that the risks associated with facilities and components will depend on the missions that they are intended to support. For example, the risks associated with a failure in a heating, ventilation, and air-conditioning (HVAC) system in a research laboratory or museum that has historical artifacts will be different from the risks associated with a failure in an HVAC system in an office building.

The mission-dependency index developed by the U.S. Navy, the U.S. Coast Guard and the National Aeronautics and Space Administration (NASA) is a process-based index that incorporates operational risk management into facilities management. A series of models and indexes developed by the U.S. Army Corps of Engineers (USACE) Engineering Research and Development Center-Construction Engineering Research Laboratory (ERDC-CERL) can be used to identify and quantify the risk of failure of building components and systems and the risk of failure of some types of infrastructure. The USACE Inland Navigation Program is an example of a different approach that can be used to quantify risks and link maintenance and repair activities to a mission.

Finding 4. Excess, underutilized, and obsolete facilities constitute a drain on the federal government's budget in costs and in forgone opportunities to invest in the maintenance and repair of mission-supportive facilities and to reduce energy use, water use, and greenhouse gas emissions.

Federal agencies report that they operate and maintain about 45,000 facilities that have become excess with respect to their missions or are underutilized. The total number of facilities that are technologically or otherwise obsolete is unknown. The cost to operate, maintain, and repair excess, underutilized, and obsolete facilities is more than $1.6 billion a year.

Private-sector organizations "have a direct incentive to dispose of unneeded facilities because they are a drain on organizational resources and are readily identifiable on their balance sheets" (NRC, 2004a, p. 97). Excess or underutilized facilities are disposed of through sales, demolition, or nonrenewal or termination of leases to free resources for other organizational requirements. In those ways, private-sector organizations manage the risk of fiscal exposure related to the ownership of facilities, reduce their maintenance and repair requirements, and reduce facilities-related expenses such as property taxes and the costs of energy, water, insurance, and security. They also manage the risk to their public image posed by abandoned and poorly maintained facilities, which could affect the public's willingness to buy their products (NRC, 2004a).

Representatives of IBM and General Motors (GM) emphasized that their organizations were unable to dispose of excess facilities effectively until the effort was managed by a central organization charged with that responsibility and the effort was funded to support the sale or demolition of excess facilities.

Two previous National Research Council studies, *Stewardship of Federal Facilities: A Proactive Strategy for Managing the Nation's Public Assets* (1998), and *Investments in Federal Facilities: Asset Management Strategies for the 21st*

Century (2004a), have made recommendations related to the disposal of excess federal facilities. *Intelligent Sustainment and Renewal of Department of Energy Facilities and Infrastructure* (NRC, 2004b) outlined a process for decision-making about when to repair, replace, renovate or dispose of facilities.

If the disposal of excess properties by federal civilian agencies can be successfully implemented, the federal government would reduce its risk of fiscal exposure related to the ownership of buildings, reduce its total operating costs, and reduce its long-term maintenance and repair requirements and costs. Disposal of excess and underutilized facilities would also help to meet other public policy objectives related to reductions in the use of energy and water, and reduction in greenhouse gas emissions. To realize those long-term savings and benefits, a coordinated, sustained, and funded effort will be required.

Finding 5. To manage and mitigate the risks posed by the ownership of facilities, high-performance private-sector organizations do the following:

- **Systematically dispose of excess and underutilized facilities.**
- **Pursue a proactive strategy to minimize their total facilities "footprint."**
- **Link maintenance and repair activities to the organization's business or mission and set priorities among them.**
- **Correlate the effects of systems-related failures with the business or mission.**
- **Correlate delays in timely maintenance and repair with sustainment cost.**

To reduce their maintenance and repair requirements, private-sector and other high-performance organizations proactively initiate strategies to minimize their total facilities footprint and the associated costs (such as the costs of equipment, furniture, and landscaping). Advances in technology allow telecommuting and other alternative work strategies that offer the potential for both public and private-sector organizations to reduce their required amounts of office or administrative space substantially, and reduce their overall maintenance and repair requirements. Some federal agencies including the General Services Administration and the Patent and Trademark Office are implementing telework strategies and reducing their need for office space. The Telework Enhancement Act of 2010 (PL 111-292) grants federal employees eligibility to telework and requires all federal agencies to establish telework policies. As the policies are implemented, there will be opportunities to reduce the overall federal facilities footprint further and to achieve substantial reductions in long-term maintenance and repair requirements and a more sustainable portfolio of federal facilities.

The primary objective of portfolio-based facilities management is to ensure that facilities-related investments enable an organization's mission. Private-sector companies, such as GM and IBM, which produce vehicles and computers, respectively, have been able to correlate the failure to invest in maintenance and repair with the mission of their organizations, which is to make a profit for their owners and shareholders. Some federal agencies—including the USACE, the Navy, NASA, and the Coast Guard—have also developed approaches for linking maintenance and repair investments to mission-related operations.

When failures do occur, some companies, such as GM, conduct a "root-cause analysis" to determine why they happened, determine the appropriate solutions, and then share the lessons learned with the managers of other facilities that may be at risk for similar failures. Root-cause analysis is a basic premise of reliability-centered maintenance, which is used by many industries, and is also being used to some degree in some federal agencies, including NASA and the Smithsonian Institution.

Delaying maintenance and repair of facilities can shorten the service lives of components and systems and ultimately result in an increase in sustainment cost. Service-life models and the Integrated Multiyear Prioritization and Analysis Tool (IMPACT) simulation model developed by ERDC-CERL can be used to quantify the costs of delaying repairs.

Finding 6. To make the outcomes of and risks posed by investments in maintenance and repair projects and activities transparent to decision-makers at all levels of the organization, facilities managers in high-performance organizations do the following:

- **Aggregate maintenance and repair requirements for some facilities' systems and components (such as life-safety systems and roofs) to provide for greater transparency and to identify operational efficiencies.**
- **Perform "knowledge-based" condition assessments; that is, tailor the frequency and level of inspection to the strategic importance of facilities and to the life cycle of systems and components to provide credible estimates of repair costs and remaining service lives.**
- **Measure outcomes as a basis of continuous improvement.**
- **Implement feedback systems to evaluate the performance of investments.**

Each of the private-sector organizations interviewed for this study identified projects that must be funded within their organizations and then fund them through sources other than the maintenance and repair account. "Must fund" projects include ones that are required for an organization to be in compliance with government regulations (such as regulations related to worker health and safety and air and water quality) because the risks associated with non-compliance—substantial fines or legal action—are much greater than the costs of the projects themselves.

To provide greater transparency in the decision-making process related to investments in maintenance and repair, some private-sector organizations group their maintenance and repair projects by component, such as roofs, HVAC, and fire protection. Grouping all the projects related to a component type makes the benefits and risks associated with projects more transparent to senior decision-makers. It also makes it easier for them to vet, screen, treat, and then set priorities among the projects that should be funded first to ensure that an organization's profits are not adversely affected by unreliable facilities systems or components.

Condition assessments that are undertaken on a multiyear cycle and that are conducted for an entire portfolio of facilities can be inefficient and expensive and the information can lose its value for decision-making relatively quickly. Some organizations are taking a "knowledge-based" approach to condition assessment. The term knowledge-based is used "to indicate that knowledge (quantifiable information) about a facility's system and component inventory is used to select the appropriate inspection type and schedule throughout a component's life cycle. Thus, inspections are planned and executed on the basis of knowledge, not merely the calendar" (Uzarski et al., 2007). Because different building systems and components have different service lives and the risks associated with the failure of some may be greater than the risks associated with the failure of others, some systems and components are inspected more often than others and at different levels of detail. By tailoring the frequency and level of inspections, a knowledge-based approach makes better use of the resources available and provides more timely and accurate data to support investment-related decisions.

Most public-sector and private-sector organizations, including federal agencies, have implemented performance measurement as a basis of continuous improvement of facilities management and other processes. Some private-sector organizations also conduct a year-end evaluation of budget performance. That type of evaluation compares the "submitted budget" the "approved budget," and the "actual funding spent" by line item. Where sizable variation exists (greater than 10 percent), the root cause of the deviation is analyzed by using the same basic process as would be used for analyzing equipment failures. The root cause of the deviation is then shared with senior decision-makers throughout the organization.

Finding 7. Investment strategies, definitions of maintenance and repair, maintenance and repair practices, and methods for budget development vary among federal agencies as a result of their different missions; the sizes, compositions, and distributions of their facilities; and their organizational cultures. The lack of common approaches makes it difficult to compare the effectiveness of maintenance and repair investments among federal agencies, to compare the benefits and pitfalls of different investment strategies, and to benchmark performance for the purpose of continuous improvement.

Between 2004 and the end of 2010, the Federal Real Property Council of the Office of Management and Budget issued guidance focused on improving the strategic management of federal buildings and structures, improving the management of the condition of facilities, developing asset management plans, implementing controls to improve the reliability of facilities-related data, and developing a set of governmentwide performance measures related to the management of portfolios of facilities (GAO, 2011a). Although a great deal of progress has been made, more is required if federal agencies are to continue to improve the management of and investment in their portfolios of facilities.

Finding 8. Reliable and appropriate data and information are essential for measuring and predicting outcomes of investments in federal facilities maintenance and repair. An array of data, tools, and technologies is available to support strategic decision-making, to quantify outcomes and risks by using empirical data, to expedite data collection, and to reduce human errors and bias.

Many factors are driving a more strategic approach to facilities management, but information tools and technologies are enabling it. Information tools and technologies are now available to monitor facilities' condition, energy use, and other performance dimensions and to collect data that can be used to reduce long-term costs, eliminate human error and bias, and increase operational efficiencies. Data and information can be the basis of higher situational awareness during decision-making, of transparency during the planning and execution of maintenance and repair activities, of an understanding of the consequences of alternative investment strategies, and of increased accountability. A wide array of tools and technologies are available to acquire and track data and to measure and predict outcomes of investments or the lack of investment.

Because the costs associated with data collection, analysis, and maintenance can be large, the committee believes that "no data before their time" should be an infrastructure-management tenet. Every system and data item should be directly related to decision-making at some level, and off-the-shelf decision-support systems should be fully integrated into decision-making processes. To the extent possible, data should be collected uniformly among federal agencies so that they will be more uniform and support the development of governmentwide performance measures and the greater use of benchmarking for agency practices and investment strategies.

Finding 9. Additional research and collaborative efforts are needed to continue to develop rapid and effective data-collection methods (such as the use of sensors and visual imaging devices), data definition and exchange standards that allow interoperability of data and software systems, and robust prediction models.

Technologies for various aspects of facilities management are continually advancing in their capabilities. Three technologies that could improve the collection and tracking of data, improve maintenance and repair activities, and provide support for decision-making are self-configuring systems, machine vision, and building information modeling (BIM).

Although the benefits of BIMs for facilities management and operations are apparent, BIM technology in its current form is best categorized as an information repository. Improved data exchange standards and software systems are needed for full interoperability of data from many systems. Interoperability, in turn, will allow more seamless integration of the data and functionalities needed to support more strategic decision-making related to maintenance and repair investments and to document outcomes.

Federal agencies participate in and support a number of efforts to develop the data and exchange standards, protocols, standard definitions, and data items that are needed if BIM is to reach its full potential as a tool for portfolio-based facilities management.

RECOMMENDATIONS

Recommendation 1 (Findings 4 and 5). To better manage the economic, physical, and environmental risks associated with facilities ownership, the federal government and its agencies should embark on a coordinated, funded, and sustained effort to dispose of excess and underutilized facilities. They should also proactively reduce their total facilities footprint through alternative work strategies and other measures.

Recommendation 2 (Findings 1, 5, and 6). Federal agencies should develop more strategic approaches for investing in facilities maintenance and repair to achieve beneficial outcomes and to mitigate risks. Such approaches should do the following:

- **Identify and set priorities among the outcomes to be achieved through maintenance and repair investments and link them to achievement of agencies' missions and other public policy objectives.**
- **Provide a systematic approach to performance measurement, analysis, and feedback.**
- **Provide for greater transparency and credibility in budget development, decision-making, and budget execution.**

Recommendation 3 (Findings 1, 2, and 3). To develop more strategic approaches to maintenance and repair investment, federal agencies should do the following:

- **Identify and set priorities among the beneficial outcomes that are to be achieved through maintenance and repair investments, preferably in the form of a 5- to 10-year plan agreed on by all levels of the orga-**

nizations. Elements of that type of plan are outlined in Chapter 7.

- Establish a risk-based process for setting priorities among annual maintenance and repair activities in the field and at the headquarters level. Guidance for doing that is contained in Chapter 7.
- Establish standard methods for gathering and updating data to provide credible, empirical information for decision support, to measure outcomes of investments in maintenance and repair, and to track and improve the results.

Recommendation 4 (Finding 6). Federal facilities program managers should plan for multiple internal and external communications when presenting maintenance and repair requests to other decision-makers and staff. The information communicated should be accurate, acknowledge uncertainties, and be available in multiple forms to meet the needs of different audiences. The basis of prediction of outcomes of a given level of investment in maintenance and repair should be transparent and available to decision-makers.

Recommendation 5 (Finding 7). Federal agencies and other appropriate organizations should continue to collaborate to develop and refine governmentwide measures for outcomes of maintenance and repair investments and to develop more standardized practices, unambiguous procedures, definitions, and models. The committee believes that those activities would be most effective if under the auspices of the Office of Management and Budget.

Recommendation 6 (Findings 6 and 8). Federal agencies should avoid the collection of data that serve no immediate mission-related purpose. Agencies should use a "knowledge-based" approach to condition assessment. Outcome metrics and models should make maximum use of existing data. When new or unique data are required to support the development of an outcome measure or model, there should be a clearly defined benefit to offset the cost of collecting and maintaining them.

Recommendation 7 (Findings 8 and 9). Federal agencies should continue to participate in and take advantage of collaborative efforts to develop rapid and effective data-collection methods (such as the use of sensors and visual imaging devices), to develop data-exchange standards that allow interoperability of data and software systems, to develop the empirical information needed for robust prediction models, and to develop practices that will reduce the cost of data collection and eliminate human error and bias.

7

Implementing a Risk-Based Strategy for Investments in Federal Facilities' Maintenance and Repair

The implementation of a more strategic, risk-based approach to investments in maintenance and repair will require changes in procedures and mindsets. It may also require a substantial investment of staff time and expertise up-front. Each agency will need to determine the most effective way to move to a risk-based investment strategy, depending on the information that it has available and its processes, resources, and culture. Once the key elements are established, however, a risk-based approach should provide for a much more effective and transparent process for decision-making about the allocation of resources for maintenance and repair activities. As the new procedures are repeated, they will become ingrained in the organizational culture and in the workforce and will require less time and effort to execute.

This chapter shows how some of the committee's recommendations could be put into action by federal facilities program managers. Topics include measures of outcomes, linking maintenance and repair investments and outcomes to a mission, guidelines for developing an annual funding request, predicting outcomes of a given level of investment in maintenance and repair, and methods for identifying risks related to deteriorating facilities.

Because missions, programs, culture, and practices vary widely among federal agencies, each agency will need to adapt the committee's guidance and examples to its own situation. It may be wise to begin with pilot projects to work through the approach before applying it to an entire organization. In addition, collaboration among agencies on pilot projects could help to identify and resolve issues more quickly and thereby help to implement risk-based approaches in all federal agencies.

MEASURES OF OUTCOMES

Outcomes of individual projects (project goals) are identified during the project planning and definition phase and can range from maintaining a heating, ventilation, and air-conditioning (HVAC) system to operate at the manufacturer's specified (or original) level of efficiency, or repairing or replacing a roof, to repairing or replacing groups of systems to achieve a higher level of efficiency (over that originally specified). Whether the project is being accomplished to conserve energy, to ensure mission capability or productivity, to lower operating costs, to improve the condition of military housing, or for any other reason, an individual project, if approached correctly, will have a defined set of outcomes. The outcomes of day-to-day activities (not on a project scale)—such as service calls, preventive maintenance (for example, lubrication and filter changes), and minor equipment replacements—are equally important because of the potential cumulative effect of neglecting them.

Identifying outcomes and the decision-making that leads to them occurs at two levels: portfolio-based (strategic) and project (tactical). Some of the outcomes identified (such as operating costs, energy use, and reliability) are more easily quantified at the portfolio level and others are more easily applied at the project level. Some (depending on the specific measure used for a given outcome) have meaning at only one of the two levels.

Regardless of how an agency goes about defining the outcomes to be achieved through its maintenance and repair program, appropriate measures are needed for planning and programming, budget development, and identifying the results of investments.

In Chapter 2, the committee identified an array of beneficial outcomes that can result from investments in maintenance and repair. All of them can be measured by using available data, technologies, and tools. Most of the measurements will be based on information that is developed after the fact (lagging measures). However, some outcomes, such as reliability and physical condition can be predicted (leading measures), that is, the outcomes of investments can be estimated before an investment is made, or before it is decided not to make the investment (the do-nothing case).

Governmentwide measures have been developed for operating costs, building condition (in the form of an index modeled on financial measures), energy use, water use, and space utilization. Deferred maintenance and repairs is also being reported, although the methods for estimating deferred maintenance and repairs vary. Government-wide measures to track greenhouse gas emissions are being developed.

In Chapter 3, the committee identified engineering-research-based indexes and models that can be used to measure the physical condition of buildings, building components, and some types of infrastructure. Risk assessment and consequence are embedded into the indexes. The indexes can also be used to predict the future physical condition of components and their remaining service lives. In doing so, they can help to identify the best time to invest in maintenance and repair so that service lives are optimized and so that systems and components can be replaced before they fail. Condition-index values can be rolled up to determine the physical condition of systems, buildings, groups of buildings or entire portfolios of buildings and infrastructure. An index to measure outcomes related to building functionality was also identified.

In this chapter, the committee identifies data sources and methods for developing measures related to reliability, accidents and injuries, building-related illnesses, claims and lawsuits, efficient operations, life-cycle costs, customer satisfaction, and public image. Because those outcomes are not now typically measured by federal agencies, they present an opportunity to collaborate to develop government-wide measures based on evidence-based empirical information. Ideally, such measures would quantify the relationships between the amount of resources invested in maintenance and repair and different levels of outcomes and risk. Such measures may require the development of models and more empirical evidence than is now available.

Table 7.1 replicates the beneficial outcomes identified in Chapter 2 and identifies related performance measures. The data, tools, and technologies that can be used to develop outcome-related measures are described in greater detail in the following section. The committee cautions that agencies should, to the extent possible, ensure that performance measures are aligned to achieve complementary objectives. Conflicts in performance measures should occur only when tradeoffs are indicated.

Measures of Mission-Related Outcomes

Reliability. Reliability of individual systems and components can be quantified as a percentage of time that they were operating in support of an agency's mission or programs. It can also be tracked as the percentage and cost of unplanned outages. Unplanned outages would need to be consistently defined but could include such events as loss of power because of faulty electrical systems, time lost when all or parts of a facility must be evacuated because of flooding from deteriorated water lines or as a result of faulty alarm systems, and the like. The number and type of unplanned outages can be measured by using data collected by computerized maintenance management systems (CMMS), building automation systems (BAS), or another asset management systems. The percentage can be measured as a ratio of hours of unplanned downtime (X hours, X days) to hours of required operating time (X hours per year, X days per year). The cost of unplanned outages could be estimated by applying an average hourly labor rate to the number of people affected and multiplying by the number of hours of downtime (this calculation could also be used to track loss of productivity). If an outage resulted in damage to equipment, research, artifacts or other property, these costs could also be quantified and added to the cost of bringing all systems back online.

TABLE 7.1 Beneficial Outcomes That Can Result from Investments in Maintenance and Repair and Outcome-Related Measures

Objectives	Outcomes	Measures
Mission-related	Improved reliability	Percentage of downtime Cost of downtime Cost of damage Service life or remaining service life
	Improved productivity	Output measures
	Functionality	Building functionality index
	Efficient space utilization	Space utilization as specified by Federal Real Property Council Cost per person (General Services Administration model)
Compliance-related	Fewer accidents and injuries	Recordable incident rate Lost time incident rate Number and cost of worker compensation claims
	Fewer building-related illnesses	Number and cost of worker compensation claims
	Fewer insurance claims, lawsuits, and regulatory violations	Number and cost of worker compensation claims Number and cost of citations or violations of regulations (such as regulations of the Occupational Safety and Health Administration)
Condition-related	Improved condition	Facility condition index (financial) Building condition indexes (physical) and other engineering-based condition indexes identified in Chapter 3
	Reduced backlog of deferred maintenance and repairs	Total cost of deferred maintenance and repairs as reported to Federal Accounting Standards Advisory Board
Efficient operations	Less reactive or unplanned maintenance	Ratio of planned maintenance to reactive maintenance
	Lower operating costs	Operating costs
	Lower life-cycle costs	Return on investment Net present value Service life extension (years)
	Cost avoidance	Net present value of maintenance and repair
	Reduced energy use	Total energy use in British thermal units (Btu) Energy intensity (Btu/sq. ft); kilowatt hours; oil equivalents (gallons)
	Reduced water use	Total gallons used Cost per gallon
	Reduced greenhouse gas emissions	Measure under development
Stakeholder-driven	Customer satisfaction	Surveys Customer service calls
	Improved public image	Surveys

This is an instance in which agencies could collaborate to define various categories of outages more specifically and "mine" their CMMS and other systems to develop empirical credible data to make the case for timely investments in maintenance and repair.

Tools and technologies to predict the reliability of equipment and systems have been developed. Reliability-centered-maintenance (RCM), which is used by private-sector organizations and by the National Aeronautics and Space Administration (NASA) and the Smithsonian Institution, takes into account the service lives of equipment and components, the probability of failure, and results. With an RCM approach, it is possible to predict the reliability of some types of equipment and components on the basis of the probability of their failure.

Reliability of building systems and components and some types of infrastructure can also be predicted from the probability of failure by using the physical condition indexes and models of service life and remaining service life described in Chapter 3.

Productivity. Loss of productivity of administrative, office, or similar types of facilities can be measured as downtime or unplanned outages, as discussed above. Predictive measures of productivity could be developed for manufacturing, some test facilities, or some operations facilities on the basis of the amount of output that can be expected if systems are 100 percent reliable. Loss of productivity could be measured as ratios of output to time and cost.

Functionality. As noted in Chapter 3, an index of the functionality of buildings and building functional areas (such as those for administration, laboratory, storage, and production) that can be used for measuring 14 categories of functionality has been developed. Some of the categories—for example, environmental life-safety, comfort, efficiency, and obsolescence—are directly related to maintenance and repair activities and investments.

Space Utilization. Most agencies are tracking space utilization but the methods for defining and calculating utilization vary. A tool that could be used to track space utilization is the cost-per-person-model (CPPM) developed by the General Services Administration (GSA). The CPPM is an Excel-based tool designed to enable users to benchmark and compute the cost per person for workspace, information technology, telecommunications, telework and other alternative work environments. It can also calculate potential cost savings for different workspace scenarios, such as those which would support telework. Additional information is available at http://www.gsa.gov/portal/content/105134.

Measures of Compliance-Related Outcomes

Accidents and Injuries. In accordance with the Occupational Safety and Health Act of 1970 as amended, and Executive Order 12196 *Occupational Safety and Health Programs for Federal Employees*, signed on February 26, 1980, federal agencies are required to track workplace accidents and submit an annual report to the Department of Labor. Federal facilities program managers in concert with the safety office could request access to this information to be able to review accident causes (such as slips, trips, and falls) and determine which ones could be prevented through maintenance and repair investments. They could then track the outcomes of maintenance and repair investments through such measures as recordable incident rate, lost-time incident rate, or by the number of related worker compensation claims or lawsuits.

The costs of accidents and injuries could be quantified by gathering data from worker compensation claims and lawsuits, if allowed by law, or by the number and cost of citations or violations of regulations (for example, violations of Occupational Safety and Health Administration standards).

Through a collaborative multiyear effort, federal agencies could potentially develop empirical information that compares the cost of maintenance activities required to prevent accidents and injuries with liability and other costs associated with accidents and injuries. Data on the costs of slips, trips, falls, and other accidents may be available from the insurance industry or from research conducted by such federal agencies as the National Institute for Occupational Safety and Health or the National Institutes of Health or from disciplines other than facilities management.

Building-Related Illnesses. Although building-related illnesses are substantially preventable through appropriate operation of building systems and components, including timely maintenance and repair activities, tracking and measuring such illnesses directly is difficult, except for major incidents, such as outbreaks of Legionnaires' disease. The costs of building-related illnesses could potentially be measured by gathering data from worker compensation claims and lawsuits, if that is allowed by law.

Building-related illnesses are closely related to indoor environmental quality (for example, temperature, humidity, ventilation rates, air particles, and water quality). Data related to those factors can be collected through building automation and energy management systems. Facilities managers should be able to cut down on building-related illnesses by gathering and carefully tracking temperature and other indoor environmental attributes to ensure that they stay within acceptable ranges according to scientific studies or state-of-the-art industry standards and through preventive maintenance activities like those identified in Chapter 2.

One indicator of potential problems related to indoor environmental quality is the type of customer service calls received. Typically, customer service calls are tracked with a CMMS. Calls related to temperature (too hot or too cold), humidity levels, moisture intrusion, air quality (odors), lack of ventilation, or water quality (tastes bad) could indicate that systems are not operating properly and require maintenance, repair, or replacement.

Measures of Condition-Related Outcomes

Most federal agencies already measure condition by using a facilities condition index (FCI) or a Condition Index (CI) as recommended by the Federal Real Property Council (FRPC). Both are lagging measures and they are based on financial data, not on the physical condition of facilities. Agencies also track total backlog of deferred maintenance and repairs and report it to the Federal Accounting Standards Advisory Board, although they use different methods for quantifying backlog.

An array of engineering-based empirically derived condition indexes for specific types of facilities and infrastructure have been developed (see Chapter 3). They can be used not only to quantify physical condition but to predict the probability of failure of building and infrastructure components on the basis of service life and remaining service life.

Measures of Outcomes Related to Efficient Operations

Measures of operating costs, energy use, and water use are already being tracked by most federal agencies as recommended by the FRPC and in accord with other federal directives. A governmentwide method for measuring greenhouse gas emissions is being developed.

Life-Cycle Cost. Life-Cycle Cost (LCC) analyses are generally not used for routine day-to-day maintenance and repair activities. However, most agency-wide maintenance and repair programs also include nonroutine large projects of which LCC analyses could be used to determine return on investment. Circular A-94 of the Office of Management and Budget, *Guidelines and Discount Rates for Benefit-Cost Analysis of Federal Programs* provides a method that could be adapted for this purpose.

Cost Avoidance. Cost avoidance results from making an investment in the near term that avoids the need for a larger investment later. One method of quantifying cost avoidance would be to analyze project scopes and develop estimates of the cost to an organization if the project is not implemented. Alternatively, failure probability analyses based on models of service life and remaining service life can be conducted. The costs of the probable failure and the costs of immediate investment can be compared on the basis of net present value.

Ratio of Planned Maintenance to Reactive Maintenance. One measure of efficient operations is the ratio of planned or programmed maintenance to reactive (unscheduled) maintenance and repair (such as emergency service calls). The ratio can be an indication of whether a facilities management organization is running smoothly through logically scheduled allocations of manpower and resources or is reacting to unexpected crisis after crisis and wasting resources through inefficient work efforts. The Association of Higher Education Facilities Officers-APPA has suggested that an appropriate ratio of planned maintenance to reactive maintenance is 75 percent or more to 25 percent or less (Rose, 2007), but there is no industry-accepted standard for the appropriate breakdowns of work. Nonetheless, in the present committee's opinion, it is safe to say that an organization that is performing more than 50 percent of its maintenance and repair on a reactive basis is not operating efficiently.

CMMS data can be used to track this measure. Comparing rates over time and by season can add definition to the measure through comparisons of similar times and weather conditions. High rates of unscheduled work could also be an indicator of deteriorating condition (which would lead to a higher rate of service calls), although they could also indicate poor workmanship, poor maintenance and repair planning, or other factors.

Measures of Stakeholder-Driven Outcomes

Customer Satisfaction. Customer satisfaction data can be tracked through on-line surveys which are conducted by many facilities management organizations. The number and type of customer service calls could also be tracked through a CMMS.

Public Image. Similar to customer satisfaction, data related to public image can be tracked through surveys of visitors to federal facilities.

LINKING MAINTENANCE AND REPAIR INVESTMENTS AND OUTCOMES TO MISSION

As noted in Chapter 1, most federal agencies have developed asset management plans that are intended to "help agencies take a more strategic approach to real property management by indicating how real property moves the agency's mission forward, outlining the agency's capital management plans, and describing how the agency plans to operate its facilities and dispose of unneeded real property, including listing current and future disposal plans" (GAO, 2011b, pp. 6-7).

The committee recommends that each agency also develop a longer-term plan for maintenance and repair investment. A longer-range plan can be used to link maintenance and repair investment clearly to organizational mission and can make maintenance and repair investment a more visible and integral component of portfolio-based facilities management. Ideally, such a plan will be developed in conjunction with and will be approved by the agency's senior executives so that there is "buy-in" from all levels of the organization. The Bureau of Overseas Buildings Operations of the U.S. Department of State has developed a longer-term (5- to 10-year) maintenance plan for its facilities portfolio that provides one example of how this could be done.

A well-developed longer-range maintenance and repair plan should provide for the following:

- Outcomes of maintenance and repair activities and investments that are aligned with the organization's mission and programs.
- A basis of communication and planning throughout the organization and

with oversight groups, including OMB and Congress.
- A framework for developing annual funding requests and budget submissions.
- Continuity in direction through organizational change and leadership turnover.

The committee recognizes that differences in agencies' missions, programs, facilities, and resources will lead to differences in the format and content of longer-range strategic maintenance plans. However, a longer-range plan should include the following basic elements:

- A clear statement of the organization's maintenance and repair investment objectives.
- An agreed-on set of outcomes related to each objective.
- Priority-setting or weighting of those outcomes.
- Identification of the types of facilities that are mission-critical or mission-supportive.
- Identification of critical types of systems and components.
- Performance goals, performance indicators, and a baseline for each outcome.
- Methods to be used for implementing maintenance and repair investments (such as preventive maintenance, recurring maintenance, and third-party financing).
- Identification of the types of risks posed by lack of timely investment.

Table 7.2 provides a hypothetical example of the items to be included in a longer-range maintenance and repair strategic plan. Guidance for developing the basic elements of a plan follows.

Step 1. Establish investment objectives and outcomes related to each objective. Five broad objectives for maintenance and repair investments were identified in Chapter 1 (shown in column 1 of Table 7.1). An array of beneficial outcomes that can be achieved and measured have also been identified (Chapters 2 and 7). Individual agencies should not expect to achieve all the identified outcomes. Rather, each agency should choose a set of outcomes that are most closely related to its investment objectives. In some cases, an agency may want to add investment objectives or categories of outcomes that are more closely related to its mission.

The emphasis should be on appropriate outcomes that are agreed to at all levels of the organization and that can be predicted, measured, defended, and verified by audit. An agency should always consider the credibility, accuracy, and value of data for developing and evaluating funding requests and for communicating with others in the agency when determining which data to collect and

TABLE 7.2 Hypothetical Example of Elements to be Included in a Longer-Range Strategic Plan for Maintenance and Repair Activities

Investment Objectives	Outcomes Related to Each Objective	Importance of Outcomes (Priority or Weighting Factor)	Mission Critical Facilities (Type or Specific Facility); Mission Supportive Facilities (Type)	Critical Systems and Components	Performance Goals, Baselines for Outcomes, and Performance Measures	Methods for Delivering Maintenance and Repair Activities	Types of Risks Posed by Lack of Investment
Enable mission	Reliability of critical systems	X percent	To be determined (TBD) by agency	Electrical systems	X percent reliability on an annual basis; to be measured by hours of unplanned outages.	Preventive maintenance	Loss of power during essential operations
Provide safe, healthy and secure workplaces	Fewer accidents and injuries	X percent	TBD by agency				
Support fiscal soundness	Lower operating costs	X percent	TBD by agency				
Operate efficiently	Improved condition	X percent	TBD by agency				
	Less unscheduled work	X percent	TBD by agency				
Support public policy goals	Reduced energy use	X percent	TBD by agency	Lighting systems	Reduce energy use by 30 percent by 2015	Third-party financing, ESPCs, and programmed major maintenance	

how best to collect them. Other considerations should include the time, effort, and cost of gathering data.

Step 2. Set priorities among the outcomes to be achieved. Each agency will need to determine which outcomes are most important to achieve and set priorities among them accordingly. One method for setting priorities is to assign weights that can be expressed as percentages. Some outcomes will be related to more than one objective and can produce multiple benefits. For example, reducing energy use may also reduce operating costs. Such relationships should be considered in the weighting process.

Final weights should not be uniformly applied without knowledge of the available resources and the demand for them. For example, a 32 percent weight for activity X may make perfect sense for a budget of $100 million. But if the budget were suddenly cut to $50 million, a 32 percent investment in activity X might produce only two-thirds of a mission-critical building. Likewise, if the budget were increased to $150 million, 32 percent might be too high, and some resources could be allocated to other projects. If weights are established in the longer-range maintenance plan, the assumptions related to the level of available resources should be clearly documented.

Step 3. Identify types of facilities or specific buildings that are mission-critical and mission-supportive. To optimize investments, agencies will need to identify the types of facilities (such as piers, museums, and hospitals) or specific buildings (such as the Pentagon) that are mission-critical or mission-supportive. Many agencies have already done that through their critical infrastructure plans, through other documents, or through the use of the mission dependency index (MDI). Such a classification will help to establish where maintenance and repair investments should be targeted to ensure that funds are being used effectively. If agencies are still targeting maintenance and repair investments to facilities that are excess, obsolete, underutilized, or slated for disposition or demolition, they should clearly indicate where and why.

Step 4. Identify critical systems and components that are most important for achieving outcomes. Agencies will need to identify the types of systems and components that are critical for achieving desired outcomes or that pose the greatest risks. As noted in Chapter 4, best-practice organizations aggregate their maintenance and repair requests by types of systems and components to create a more transparent linkage to specific building performance. Aggregating requests that way also allows decision-makers to understand more easily the relative importance of systems and components for various investment outcomes. Agencies that use the MDI can extrapolate critical systems and components from it. The Army's Engineering Research and Development Center-Construction Engineering

Research Laboratory has also created a research-based Component Importance Index that can be used to identify critical components (Uzarski et al, 2007).

Critical systems and components would likely include the following:

- Enclosures—façades, windows, and doors,
- Roofs,
- Heating, ventilation, and air-conditioning (HVAC),
- Lighting,
- Electrical distribution,
- Fire protection systems,
- Security systems,
- Plumbing and water fixtures,
- Roadways, parking, and paving,
- Industrial type systems—cranes, conveyors, and the like.

Step 5. Establish performance goals, baselines for outcomes, and performance measures. Establishing performance goals, baselines for outcomes, and performance measures is essential for tracking the effectiveness of maintenance and repair investments, for providing feedback on progress, and for indicating where investment objectives, outcomes, or procedures require adjustment. "Buy-in" at all levels of the organization is needed if sustained progress is to be achieved.

Step 6. Identify the primary methods to be used for delivering maintenance and repair activities. Maintenance and repair activities can be delivered through programs for preventive maintenance, programmed major maintenance, replacement, or in some cases, public-private partnerships or third-party financing (such as through energy savings performance contracts). Identification of the methods of delivery will help agencies to determine the level of resources that should be allocated to each type of maintenance activity and to repair projects and to determine when repair projects can be funded through methods other than direct appropriations.

Step 7. Identify the types of risks posed by lack of timely investment. Identifying the types of risks posed by not investing in deteriorating facilities, systems, and components is important for providing more transparency in the decision-making process and for communicating with staff at all organizational levels. For a longer-range maintenance plan, a general description of the types of risks, as opposed to the level or quantification of risks, will be appropriate because risks may change every year or more often. In all cases, the description of risks should be credible. Methods for identifying risks related to deteriorating facilities, systems, and components are described later in this chapter.

GUIDELINES FOR DEVELOPING AN ANNUAL
RISK-BASED FUNDING REQUEST

In any given year, the number of required maintenance activities and repair projects will exceed available funding. A longer-range maintenance and repair strategic plan can provide the framework for determining the types of activities and projects that are the most critical to fund for a sustained period. Determining the level of funding for maintenance activities and identifying specific repair projects that should be funded in a given budget year require a more detailed analysis—one that still recognizes that budget requests are generally developed 2 years in advance of funding.

Table 7.3 lists the types of elements that should be identified in annual funding requests. A standard template could be developed and then used by facilities managers at the field level and rolled up to headquarters. The headquarters staff can use the same template to reset priorities among projects across an agency to align with organizational objectives and to present a unified request to decision-makers in the agency.

Using the same template at all levels of the organization will help to embed new processes, provide for more consistent communication and messages, and provide transparency about how budget submissions are being developed.

Step 1: Categorize identified maintenance activities and repair projects in an organizational framework for investment. Facilities program managers at the field level or at headquarters should group all their identified maintenance activities and repair projects by categories of critical systems and components and by whether they are mission-critical or mission-supportive facilities as identified in longer-range maintenance plan (if available). At the field level, it should be possible to identify the specific facility or groups of facilities where the maintenance activities and repair projects will be implemented.

Step 2. Determine the cost of the maintenance activities and repair projects and identify the method of delivery. The costs of maintenance activities and repair projects can be verified through parametric estimates, estimates by agency experts, collection of estimates from subordinate organizations, knowledge-based condition assessments, or any other method that facilities program managers might use that has credibility in the organization. The methods to be used for executing maintenance activities or repair projects (such as programmed major maintenance or third-party financing) should also be identified.

Step 3. Identify the outcomes to be achieved. This can be done in two phases. First, list the outcomes specifically identified in the longer-range maintenance plan that have highest priority in the organization. Second, identify other credible outcomes for a specific project that could also have an organizational benefit that is not called out in the longer-range maintenance plan. That can take the form of

TABLE 7.3 Hypothetical Template for Setting Priorities Among Maintenance and Repair Activities to Be Included in an Annual Funding Request

Projects Classified by Critical Systems and Components	Mission-Critical Facilities	Mission-Supportive Facilities	Costs of Projects	Method of Delivery	Outcomes to Be Achieved	Other Potential Benefits	Type and Level of Potential Risks If Not Funded	Risk Ratings
By category established in longer-range maintenance strategic plan	By category established in longer-range maintenance strategic plan	By category established by longer-range maintenance strategic plan	Determined at time of funding request (conduct knowledge-based condition assessments where appropriate)	Preventive maintenance programmed major maintenance; energy savings performance contract; public-private partnership	All the outcomes that apply as established in longer-range maintenance strategic plan	Additional information required for well-informed decision making	Narrative with supporting quantitative data	As calculated using CRR or other process (see below)

a narrative with backup evidence-based information that can be verified. The process should ensure that critical, credible information is available for well-informed decision-making on behalf of the entire organization.

Step 4. Identify the type and level of risks incurred if the maintenance activities and repair projects are not funded in the relevant fiscal year. This step is intended to ensure that the most critical requirements rise to the top of the funding requests and that senior decision-makers understand the implications of not funding maintenance activities or particular repair projects in the relevant fiscal year. It is also intended to provide greater transparency, credibility, and accountability in budget formulation and execution.

Step 5. Setting priorities among projects. A variety of methods are available for ranking all the proposed repair projects and setting priorities among maintenance and repair activities. They include the Analytic Hierarchy Method and the Delphi Method.

The Analytic Hierarchy Method (ASTM 1765-07e1) allows consideration of multiple decision-making criteria in the priority-setting process. The multiple ranking criteria are weighted through pairwise comparisons, and the relative importance of each criterion becomes established. Through this process, the various decision-making criteria are weighted to provide an objective measure of the priority of a specific activity or project.

The Delphi Method (Linstone and Turoff, 1975) is another approach whereby a multiple-stage protocol is used to obtain a consensus expert opinion. Experts are asked to respond to questions, and after each stage a facilitator summarizes the results. Eventually, with revision of responses, the range of responses decreases and the group as a whole converges toward a consensus. Typically, the process has guidelines about what constitutes a consensus and about the number of rounds. The method can be applied with face-to-face meetings or questionnaires.

Whatever process is used, it should be documented and used consistently by the various field offices to ensure that when a request is sent to headquarters, it is credible and easily communicated.

PREDICTING OUTCOMES OF A GIVEN LEVEL OF INVESTMENT IN MAINTENANCE AND REPAIR

Depending on the outcomes selected in Step 3, one or more applicable prediction models (see Chapter 3) should be used to create projects and develop priorities for programs at the field level. Additionally, a consequence (what-if) analysis should be made that considers different possible investment levels (such as likely, lower limit, and upper limit).

After the field-level requests have been submitted to the facilities management office at headquarters, the headquarters staff will need to reset priorities among the requests to meet overall organizational objectives. They should also roll up the predicted outcomes of maintenance activities and repair projects to quantify the expected results on a portfolio-wide basis to the extent possible (for example, total energy reductions across all facilities). Once the overall funding request is prepared, the headquarters staff can use a funding "cutoff" line (such as $3 million or $5 million) to show which activities and projects can be funded at a given level of investment and which ones cannot. Performing a consequence analysis through the use of applicable prediction models can change the project mix to maximize desired outcomes. A funding request in this type of format will provide greater transparency about the repair projects that are considered to have highest priority, their costs, and the benefits that the organization can expect. It will also make clear the risks posed by not funding projects.

METHODS FOR IDENTIFYING RISKS RELATED TO DETERIORATING FACILITIES

The risk-analysis literature offers multiple entry points into answering the following questions:

1. What can go wrong?
2. What are the chances that something with serious consequences will go wrong?
3. What are the consequences if something does go wrong?
4. What can be done and what options are available? How can the consequences be prevented or reduced?
5. What are the associated tradeoffs in costs, benefits, and risks? How can recovery be enhanced if the scenario occurs?
6. What are the effects of current management decisions on future options? How can key local officials, expert staff, and the public be informed to reduce concern and increase trust and confidence?

Some entry points are entirely quantitative and others mix quantitative and qualitative data. Given the reality that federal agencies have offices and properties around the world, a risk-based approach to investment is most logically implemented in two phases: screening to set priorities among the maintenance activities and repair projects and then detailed analyses from among those chosen as having high priority. The information developed during the overall screening process can be used to identify types of risks and can be used in the longer-range maintenance plan. More detailed analyses will be more appropriate for the development of annual funding requests.

It is essential first to identify the vulnerabilities of federal facilities, systems and components, and then evaluate the vulnerabilities in the context of importance to mission fulfillment. The mission dependency index and the USACE's asset management strategy (see Chapter 4) incorporate risk-based screening processes to determine which facilities and components are the most critical for an agency's mission and the failure of which poses the greatest risks to operations and mission achievement.

Two additional examples identified by the committee clearly illustrate the screening process and are based on a combination of science, engineering, and legibility. Both examples deal with vulnerability to terrorism but the principles of how to set priorities are transferable to building and infrastructure failures.

Apostolakis and Lemon (2005) developed a screening model to identify vulnerabilities of a university-centered community on a single campus. The authors rated asset vulnerability on a continuum that began with red (most vulnerable) and proceeded through orange, yellow, blue, to green (least vulnerable). Then they studied how the elements of the infrastructure—such as natural gas, water, and electricity—were interconnected. Next, they developed a "value tree" that reflected the values and perceptions of the decision-makers and other important stakeholders about each asset. The value categories included health, safety and environmental effects; economic effects on property, academic-institution operations; stakeholder effects; and effects on public image.

The values were then weighted. The greatest weights were assigned to effects on people, followed by effects on the environment, university programs, and so on. The vulnerability and value data were then connected to produce a priority list of campus projects that the university could act on. Those projects ranged from welding manhole covers to building independent infrastructure supply lines to adding backup components.

Leung, Lambert, and Mosenthal (2004) built a screening tool to set priorities for investments to protect bridges in Virginia. Their analysis was more complex than the first one in that it considered multiple major assets in different unconnected locations as well as specific singular assets, but the logic was the same. Scenarios that could degrade the system were identified, ranked according to their potential adverse events, and then compared with the system's existing resilience, robustness, and redundancy. At every step, analysts integrated historical data and expert judgment. After completing the initial risk assessment, they gathered information on the cost, on engineering feasibility, and on policy options. The security, economic, and safety implications of options were then studied at the national, regional, and local levels. Simple decision trees were built to aid decision-makers in understanding the options before priorities were set.

Although the first method was applied to a single area and the second was applied at regional and national levels, both followed risk-analysis principles, including identifying critical assets, examining their vulnerability, and setting priorities for their protection.

For individual facilities, systems, and components, traditional engineering approaches can be applied to priority projects. It requires setting numerical performance measures in a risk-related framework (see Ellingwood, 1994). In civil or structural engineering, reliability is defined with a "reliability index" or a "safety index" (see Ang and Cornell, 1974), which is related explicitly to an underlying probability of performance. For example, in designing a building, structural engineers often apply numerical "factors of safety" or "load and resistance factors" to ensure the safety of the building against external loads, including earthquakes or windstorms. That design procedure can be integrated with estimates of probability of failure. Indeed, engineering standards for the design of buildings and bridges are now widely based on this approach (see ASCE, 2005; and ASCE/ANSI, 2006).

Inspection and repair intervals may be optimized or nearly optimized by decision-makers in order to maintain reliability of function. Engineered components and systems deteriorate with time and use. To maintain a given level of reliability (probability of performance), inspection and repair of the critical components at prescribed time intervals are necessary. For example, to maintain the performance of a bridge component against fatigue failure, the interval of inspection and repair can be altered to ensure that a specified reliability (probability of nonfailure) is maintained (see discussion of knowledge-based condition assessments in Chapter 3).

An important but sometimes overlooked aspect of risk analysis is that of probability of occurrence. For example, the consequences (such as lives lost, dollar value, or mission interruption) of an adverse event may be very high, but if the probability of the event is extremely small, the risk will be minimal. (See Appendix C for further discussion.)

At the individual asset level, a critical complication is uncertainty and how it affects performance (Ang and Tang, 2007; Frangopol et al., 2001). It may be that a critical component should perform flawlessly for 5 years. However, some perform beyond expectations and others fail far more rapidly than expected. Consequently, it is essential that facilities program managers use knowledge-based inspection practices and set inspection, maintenance and repair schedules that recognize the reality that some critical components of important assets will fail before they are expected to.

The committee recognizes that many federal agencies will not have the resources to undertake detailed engineering-based analyses for the majority of maintenance activities and repair projects that they must evaluate annually. One method of analysis and priority-setting that could potentially be used by agencies involves the use of risk-rating charts developed for reliability-centered maintenance (RCM) processes. The process is relatively simple and does not require the collection of large amounts of data, but it does require knowledgeable, experienced facilities management professionals.

In this type of process, risk ratings are established for specific types of components (for example, roofs, HVAC systems, and some equipment) and for subcategories of them. Each risk rating for a specific component includes two primary elements of risk: probability of failure (POF), and failure consequence (FC). The component risk rating (CRR) is the product of POF and FC, or CRR = POF × FC (Figure 7.1).

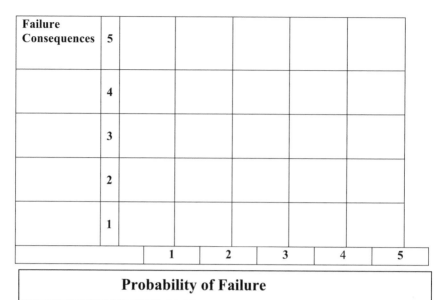

FIGURE 7.1 Component risk rating chart.

Some organizations have standard facility categories.[1] In addition, each agency might have several mission-specific categories of components. For example, runways, air traffic control towers, and airplane hangars are mission-specific for the Air Force, piers and cargo loading cranes are mission-specific for the Navy, and museums are mission-specific for the Smithsonian Institution.

Assigning Probability of Failure Ratings. Typical POF ratings are shown below:

- *5*—The probability of a failure in the given fiscal year is high.
- *3*—The probability of a failure in the given fiscal year is moderate.
- *1*—The probability of a failure in the given fiscal year is low.
- *2 and 4*—Variances between the other three ratings as determined by expert opinion.

---

[1]Standard facility asset or component groupings for an agency can follow the *ASTM Uniformat II* classification approach (ASTM E-1557) or any other standardized approach recognized or used by the agency. *Uniformat II* recognizes 17 building systems. The Department of Defense and other agencies have methodologies for grouping facilities by importance (mission-specific) or use (category codes). It is not the intent of the present committee to suggest that agencies "reinvent" groupings, rather they should use a logical, reasonably comprehensive approach that is compatible with their facilities management approach and regulations.

The literature includes many ways of quantifying risk that are based on how people describe it. See, for example, Reagan et al., 1989.

Assigning Failure Consequence Ratings. Examples of typical FC ratings are shown below:

- *5*—Serious consequences, such as death, injury, illness, extended shut-down of an agency's mission, substantial costs, substantial environmental effects, or noncompliance with regulations.
- *3*—Moderate consequences, such as reduced comfort, increase in long-term ownership costs, or delay in mission-completion date by some number of days or weeks.
- *1*—Minimal consequences.
- *2 and 4*—Variances between the other three ratings as determined by expert opinion.

Example 1: With HVAC as a component category, the subcategories (defined as a type of facilities that have similar risk characteristics) are defined and a risk rating is assigned:

- *Warehouse ventilation where multiple air-supply units supply air to the same large space.* Failure of a single unit will have a small effect on mission-related operations and the FC rating would be 1.
- *Laboratory air supply in which air cleanliness and temperature and humidity are critical for accurate results.* Laboratories are usually of such a size that only one HVAC unit supplies a specific laboratory and the FC rating for this subcategory of components might be 4 or 5.

Example 2: With roofs as a component category, a specific agency might identify the subcategories and assign a risk rating as follows:

- *Roofs on aircraft hangars.* If a serious leak occurs in the roof of an aircraft hanger, it may have a small effect on the assets within (airplanes and related components), because such assets are designed to withstand the elements. The FC rating might be 1.
- *Roofs on central data centers.* Roof leaks on data centers could shut down an agency's operations for an extended period and the FC rating might be 5.

Once the POF and FC ratings are established for each component category and for the subcategories, the component risk rating (CRR) can be calculated.

In the committee's experience, an agency would typically have no more than 20 categories of critical components and perhaps 10 subcategories of facility types with an average of about two risk ratings each. That would mean establishing about 60 CRRs per agency. Once the CRRs ratings have been established by the facility management program, they should be reviewed by senior-decision-makers to ensure "buy-in" at all levels of the organization.

The CRRs can now be used for all facility components and types. Although the task of establishing the CRRs will require an investment of time and effort upfront, established CRRs can be used in future years with little additional effort.

References

AGC (Associated General Contractors of America). 2006. The Contractors' Guide to BIM. Washington, D.C.: AGC.

Allen, Grahame. 2001. The Private Finance Initiative (PFI). London: House of Commons Library, Economic Policy and Statistics Section.

Ang, A.H-S., and C.A. Cornell. 1974. Reliability bases of structural safety and design. Journal of Structural Design 100(9):1755-1769.

Ang, A.H-S., and W.H. Tang. 2007. Probability Concepts in Engineering: Emphasis on Applications to Civil and Environmental Engineering. 2nd ed. New York: John Wiley and Sons.

Antelman, A., J. Dempsey, and W. Brodt. 2008. Mission Dependency Index—A metric for determining infrastructure criticality. Pp. 141-146 in Infrastructure Reporting and Asset Management: Best Practices and Opportunities. Reston, Va.: American Society of Civil Engineers.

Apostalakis, G., and D. Lemon. 2005. A screening methodology for the identification and ranking of infrastructure vulnerabilities due to terrorism. Risk Analysis 25:361-376.

APWA (American Public Works Association). 1992. Plan. Predict. Prevent. How to Reinvest in Public Buildings. Special Report No. 62. Chicago, Ill.: APWA.

ASCE (American Society of Civil Engineers). 2005. Seismic Design Criteria for Structures, Systems, and Components in Nuclear Facilities. Standard 43-05. Reston, Va.: ASCE.

ASCE and ANSI (American National Standards Institute). 2006. Guidelines for the Seismic Rehabilitation of Buildings. Reston, Va.: ASCE.

ASHRAE (American Society of Heating, Refrigeration and Air-Conditioning Engineers). 2009. 2009 ASHRAE Handbook—Fundamentals. Atlanta, Ga.: ASHRAE.

Ballesty, S., J. Mitchell, R. Drogemuller, H. Schevers, C. Linning, G. Singh, and D. Marchant. 2007. Adopting BIM for Facilities Management: Solutions for Managing the Sydney Opera. Available at http://eprints.qut.edu.au/27582/. Accessed April 20, 2011.

Bettinghaus, E.P., and M.J. Cody. 1994. Persuasive Communications. Fort Worth, Tex.: Harcourt, Brace, Jovanovich College Publishers.

Boyce, P.R., C. Hunter, and O. Howlett. 2003. The Benefits of Daylighting Through Windows. Troy, N.Y.: Lighting Research Center.

Brucker, B.A., M.P. Case, W. East, B. Huston, S. Nachtigall, J. Shockley, S. Spangler, and J. Wilson. 2010. Building Information Modeling (BIM)—A Roadmap for Implementation to Support MILCON Transformation and Civil Works Projects. Engineering Research and Development Center Technical Report. ERDC TR-06-10. Available at http://www.cecer.army.mil/techreports/ERDC_TR-06-10/ERDC_TR-06-10.pdf. Accessed April 20, 2011.

Campbell, D. 2011. USPTO Telework. Presentation at Forum on Alternative Work Arrangements, Keck Center of the National Academies, Washington, D.C., June 9, 2010.

Dempsey, J. 2009. A Coast Guard pilot to make better facility decisions. Journal of Building Information Modeling, Fall, p. 26.

DOD (Department of Defense). 2001. Report to Congress: Identification of the Requirements to Reduce the Backlog of Maintenance and Repair of Defense Facilities. Washington, D.C.: DOD.

Du, Z., and X. Jin. 2007. Detection and diagnosis for sensor fault in HVAC systems. Energy Conservation and Management 48(3):693-702.

Eastman, C., P. Teicholz, R. Sacks, and K. Liston. 2008. BIM Handbook: A Guide to Building Information Modeling for Owners, Managers, Designers, Engineers and Contractors. Hoboken, New Jersey: John Wiley & Sons, Inc.

Ellingwood, B.R. 1994. Probability-based codified design: Past accomplishments and future challenges. Journal of Structural Safety 13(3):159-176.

Ellis, M.W., and E.H. Mathews. 2002. Needs and trends in building and HVAC system design tools. Building and Environment 37(5):461-470.

FEMP (Federal Energy Management Program). 2010. Annual Report to Congress on Federal Government Energy Management and Conservation Programs Fiscal Year 2007. Available at http://www1.eere.energy.gov/femp/pdfs/annrep07.pdf. Accessed April 28, 2011.

Fernandez, N., M. Brambley, S. Katipamula, H. Cho, J. Goddard, and L. Dinh. 2010. Self-Correcting HVAC Controls Project Final Report. PNNL-19074. Richland, Wash.: Pacific Northwest National Laboratory.

FFC (Federal Facilities Council). 1996. Budgeting for Facilities Maintenance and Repair. Technical Report No. 131. Washington, D.C.: National Academy Press.

FFC. 2005. Implementing Health-Protective Features and Practices in Buildings. Washington, D.C.: The National Academies Press.

Frangopol, D.M, J.S. Kong, and E.S. Gharaibeh. 2001. Reliability-based life-cycle management of highway bridges. Journal of Computing in Civil Engineering 15(1):27-35.

GAO (Government Accountability Office). 2003. High Risk Series. Federal Real Property. GAO-03-122. Washington, D.C.: GAO.

GAO. 2008. Federal Real Property: Government's Fiscal Exposure from Repair and Maintenance Backlogs Is Unclear. Washington, D.C.: GAO.

GAO. 2011a. Federal Real Property: An Update on High Risk Issues. Washington, D.C.: GAO.

GAO. 2011b. Federal Real Property: The Government Faces Challenges to Disposing of Unneeded Buildings. GAO11-370T. Washington, D.C.: GAO.

GAO. 2011c. The Federal Government's Long-Term Fiscal Outlook. January 2011 Update. Washington, D.C.: GAO. Available at http://www.gao.gov/new.items/d11451sp.pdf. Accessed June 30, 2011.

Gillespie, T.D., M.W. Sayers, and C.A.V. Queiroz. 1986. The International Road Roughness Experiment: Establishing Correlation and Calibration Standard for Measurement. Technical Report No. 45. Washington, D.C.: The World Bank.

Golabi, K., and R. Shepard. 1997. Pontis: A system for maintenance optimization and improvement of U.S. bridge networks. Interfaces 27(1):71-88.

Goodman, E. 2009. A tone-deaf message on mammograms. Boston Globe. November 27.

Greenberg, M. 2009. Risk analysis and port security: Some contextual observations and considerations. Annals of Operations Research 187(1):121-136.

Grussing, M.N., D.R. Uzarski, and L.R. Marrano. 2006. Condition and Reliability Prediction Models Using the Weibull Probability Distribution. Pp. 19-24 in Proceedings of the Ninth International Conference Applications of Advanced Technologies in Transportation. Reston, Va.: American Society of Civil Engineers. August.

Grussing, M.N., D.R. Uzarski, and L.R. Marrano. 2009. Building infrastructure functional capacity measurement framework. ASCE Journal of Infrastructure Systems, December, pp. 371-377.

GSA (General Services Administration). 2006. Fiscal Year 2005 Federal Real Property Report. Washington, D.C.: GSA.

GSA. 2009. Fiscal Year 2008 Federal Real Property Report. Washington, D.C.: GSA. Available at http://www.gsa.gov/graphics/ogp/FY_2008_Real_Property_Report.pdf. Accessed February 28, 2011.

GSA. 2010. Fiscal Year 2009 Federal Real Property Report. Washington, D.C.: GSA. Available at http://www.gsa.gov/graphics/ogp/FY2009_FRPR_Statistics.pdf. Accessed February 28, 2011.

Haimes, Y.Y. 1991. Total risk management. Risk Analysis 11(2):169-171.

Indiana University. 2010. BIM Standards. Available at http://www.indiana.edu/~uao/iubim.html. Accessed April 4, 2011.

Jones, W., and R. Bukowski. 2001. Critical Information for First Responders, Whenever and Wherever It Is Needed. Gaithersburg, Md.: National Institute of Standards and Technology. Available at http://fire.nist.gov/bfrlpubs/fire01/PDF/f01144.pdf. Accessed July 21, 2010.

Jordani, D. 2010. BIM and FM: The portal to lifecycle facility management. Journal of Building Information Modeling, Spring, pp. 13-16.

Kaplan, S., and B.J. Garrick. 1981. On the quantitative definition of risk. Risk Analysis 1(1):11-27.

Keston Institute (Keston Institute for Public Finance and Infrastructure Policy). 2011. An Informational Summary on Public Private Partnerships. Available at http://ncsu.edu/iei/programs/growth/resources/PPPPrimer.pdf. Accessed June 21, 2011.

Kidd, R. 2010. Statement of Richard Kidd, Program Manager, Federal Energy Management Program, Office of Energy Efficiency and Renewable Energy, U.S. Department of Energy, Before the Subcommittee on Federal Financial Management, Government Information, Federal Services and International Security, Committee on Homeland Security and Government Affairs, U.S. Senate. January 27, 2010. Available at http://hsgac.senate.gov/public/index.cfm?FuseAction=Hearings.Hearing&Hearing_ID=c7cb1779-8aa1-4250-8dfe-18e06b579af1. Accessed February 28, 2011.

Klass, P. 2009. Fearing a flu vaccine—And wanting more of it. New York Times. November 10.

Kolata, G. 2009. Get a mammogram. No don't. Repeat. New York Times. November 21.

Koren, G., and N. Klein. 1991. Bias against negative studies in newspaper reports of medical research. Journal of the American Medical Association 266:1824-1826.

Laster, S., and A. Olatunji. 2007. Autonomic computing: Towards a self-healing system. Pp. 62-78 in Proceedings of the Spring 2007 American Society for Engineering Education Illinois-Indiana Section Conference. Washington, D.C.: American Society for Engineering Education.

Leung, M., J.H. Lambert, and A. Mosenthal. 2004. A risk-based approach to setting priorities in protecting bridges against terrorist attacks. Risk Analysis 24(4):963-984.

Linstone, H., and M. Turoff. 1975. The Delphi Method: Techniques and Applications. Reading, Mass.: Addison-Wesley.

Liu, X., and B. Akinci. 2009. Requirements and evaluation of standards for integration of sensor data with building information models. Proceedings of the 2009 ASCE International Workshop on Computing in Civil Engineering. Austin, Tex.: American Society of Civil Engineers.

Lowrance, W.W. 1976. Of Acceptable Risk. Los Altos, Calif.: William Kaufmann.

Lufkin, P., A. Desai, and J. Janke. 2005. Estimating the restoration and modernization costs of infrastructure and facilities. Public Works Management and Policy, July, pp. 40-52.

McNabb, W. 2010. GM Asset Management. Presentation to the NRC Committee on Predicting Outcomes of Investments in Maintenance and Repair for Federal Facilities, February 18, 2010, Washington, D.C.

Menzel, K., and D. Pesch. 2008. Towards a wireless sensor platform for energy efficient building operation. Tshinghua Science and Technology 13:381-386.

Moran, J. 2010. NNSA Roof Asset Management Program. Presentation to Federal Facilities Council, October 27, Washington, D.C.

Muto T., M. Tomita, S. Kikuchi, and T. Watanabi. 1997. Methods to persuade higher management to invest in health promotion programs in the workplace. Occupational Medicine 47(4):210-216.

NACUBO (National Association of College and University Business Officers). 1991. Managing the Facilities Portfolio: A Practical Approach to Institutional Facility Renewal and Deferred Maintenance. Washington, D.C.: NACUBO.

NAPA and NRC (National Academy of Public Administration and National Research Council). 2010. Choosing the Nation's Fiscal Future. Washington, D.C.: The National Academies Press.

NAS-NAE-NRC (National Academy of Sciences-National Academy of Engineering-National Research Council). 2008. What You Need to Know About Energy. Washington, D.C.: The National Academies Press.

NASA (National Aeronautics and Space Administration). 1996. Reliability Centered Maintenance Guide for Facilities and Collateral Equipment. Washington, D.C.: NASA.

NIBS (National Institute of Building Sciences). 2008. Assessment to the U.S. Congress and U.S. Department of Energy on High Performance Buildings. Washington, D.C.: NIBS.

NRC (National Research Council). 1990. Committing to the Cost of Ownership: Maintenance and Repair of Public Buildings. Washington, D.C.: National Academy Press.

NRC. 1993. The Fourth Dimension in Buildings: Strategies for Minimizing Obsolescence. Washington, D.C.: National Academy Press.

NRC. 1998. Stewardship of Federal Facilities: A Proactive Strategy for Managing the Nation's Public Assets. Washington, D.C.: National Academy Press.

NRC. 2004a. Investments in Federal Facilities: Asset Management Strategies for the 21st Century. Washington, D.C.: The National Academies Press.

NRC. 2004b. Intelligent Sustainment and Renewal of Department of Energy Facilities and Infrastructure. Washington, D.C.: The National Academies Press.

NRC. 2008. Core Competencies for Federal Facilities Asset Management Through 2020. Washington, D.C.: The National Academies Press.

NRC. 2009. Science and Decisions: Advancing Risk Assessment. Washington, D.C.: The National Academies Press.

NRC. 2011. Achieving High-Performance Federal Facilities: Strategies and Approaches for Transformational Change. Washington, D.C.: The National Academies Press.

NSTC (National Science and Technology Council). 2008. Federal Research and Development Agenda for Net-Zero Energy, High-Performance Green Buildings. Washington, D.C.: NSTC.

Orringer, O. 1990. Control of Rail Integrity by Self-Adaptive Scheduling of Rail Tests. DOT/FRA/ORD-90/05. Washington, D.C.: U.S. Department of Transportation, Federal Railroad Administration, Office of Research and Development.

Parashar, M., and S. Hariri. 2005. Autonomic computing: An overview. Lecture Notes in Computer Science 3566:257-269.

Reagan, R., F. Mosteller, and C. Yountz. 1989. Quantitative meanings of verbal probability expressions. Journal of Applied Psychology 74(3):433-442.

Resendiz, E.Y., L.F. Molina, J.M. Hart, J.R. Edwards, S. Sawadisavi, N. Ahuja, and C.P.L. Barkan. 2010. Development of a machine-vision system for inspection of railway track components. Paper presented at the 12th World Conference on Transport Research, Lisbon, Portugal. Available at http://intranet.imet.gr/Portals/0/UsefulDocuments/documents/03355.pdf.

Roberti, M. 2009. Bar-code technology is not cheaper than RFID. RFID Journal. June 29. Available at http://www.rfidjournal.com/article/view/5005.

Rose, R. 2007. Buildings: The Gift That Keep on Taking: A Framework for Decision Making. Alexandria, Va.: Center for Facilities Research-APPA.

Sallans, B., D. Bruckner, and G. Russ. 2006. Statistical Detection of Alarm Conditions in Building Automation Systems. 2006 IEEE International Conference on Industrial Informatics. Singapore: IEEE.

Schlake, B.W., S. Todorovic, J.R. Edwards, J.M. Hart, N. Ahuja, and C.P.L Barkan. 2010. Machine vision condition monitoring of heavy-axle load railcar structural underframe components. Proceedings of the Institution of Mechanical Engineers, Part F: Journal of Rail and Rapid Transit.

Shahin, M.Y. 2005. Pavement Management for Airfields, Roads, and Parking Lots, 2nd ed. Dordrecht, The Netherlands: Springer Science + Business Media.

Shahin, M.Y., and S.D. Kohn. 1979. Development of a Pavement Condition Rating Procedure for Roads, Streets, and Parking Lots, Vol. 1: Condition Rating Procedure. Technical Report M-268. Champaign, Ill.: U.S. Army Corps of Engineers Construction Engineering Research Laboratory. July.

Shahin, M.Y., M.I. Darter, and S.D. Kohn. 1976. Development of a Pavement Maintenance Management System. Vol. 1: Airfield Pavement Condition Rating. AFCEC-TR-76-27. Champaign, Ill.: U.S. Army Corps of Engineers Construction Engineering Research Laboratory. Available at http://trid.trb.org/view.aspx?id=56392.

Shahin, M.Y., D.M. Bailey, and D.E. Brotherson. 1987. Membrane and Flashing Condition Indexes for Built-Up Roofs, Vol. 1: Development of the Procedure. Technical Report M-87/13/ADA190367. Champaign, Ill.: U.S. Army Corps of Engineers Construction Engineering Research Laboratory.

Shih, G. 2009. Game changer in retailing, Barcode is 35. New York Times. June 25.

Siegrist, M., and G. Cvetkovich. 2001. Better negative than positive? Evidence of a bias for negative information about possible health dangers. Risk Analysis 21:199-206.

Smith D. 2007. An introduction to BIM. Journal of Building Information Modeling, Fall, pp. 12-14.

St. Thomas, R. 2010. IBM Facilities M&R. Presentation to the National Research Council, Washington, D.C., February 18.

TRB (Transportation Research Board). 2010. Performance Specifications for Rapid Renewal. Phase 2 Specifications and Development. Washington, D.C.: The National Academies Press.

Tsai, Y.-C., K. Vivek, and R. Merserau. 2010. Critical assessment of pavement distress segmentation methods. ASCE Journal of Transportation Engineering 136(1)11-19.

Union Pacific. 2005. Union Pacific unveils $8.5 million state-of-the-art track inspection vehicle. Available at http://www.uprr.com/newsinfo/releases/capital_investment/2005/1216_ec5.shtml. Accessed September 25, 2011.

Urban Institute. 1994. Issues in Deferred Maintenance. Washington, D.C.: Urban Institute.

USACE (U.S. Army Corps of Engineers). 2010. PIANC WG 129 Comparison of National Practice and Standards. Draft of Chapter 3. Champaign, Ill.: U.S. Army Corps of Engineers Construction Engineering Research Laboratory.

USPTO (U.S. Patent and Trademark Office). 2010. Introduction: Telework at the United States Patent and Trademark Office. Available at http://www.uspto.gov/about/offices/cao/TeleworkAnnual09FINAL_Section_508.pdf. Accessed May 15, 2011.

Uzarski, D.R. 2004. Knowledge-Based Condition Assessment Manual for Building Component-Sections. Champaign, Ill.: U.S. Army Corps of Engineers Construction Engineering Research Laboratory.

Uzarski, D.R. 2006. Deficiency vs. distress-based inspection and asset management approaches: A primer. APWA Reporter 73(6):50-52.

Uzarski, D.R., and L.A. Burley. 1997. Assessing building condition by the use of condition indexes. Pp. 365-374 in Proceedings of the ASCE Specialty Conference Infrastructure Condition Assessment: Art, Science, Practice. Reston, Va. August.

Uzarski, D.R., M.I. Darter, and M.R. Thompson. 1993. Development of Condition Indexes for Low-Volume Railroad Trackage. Transportation Research Record (Transportation Research Board, National Research Council) 1381, pp. 42-52.

Uzarski, D.R., N.M. Grussing, and J.B. Clayton. 2007. Knowledge-based condition survey inspection concepts. ASCE Journal of Infrastructure Systems 13(1):72-79.

Wald, M. 2009. Smart electric utility meters, intended to create savings, instead prompt revolt. New York Times. Pp. 1A and 16A. December 14.

Weber, E., N. Siebenmorgen, and M. Weber. 2005. Communicating asset risk: How recognition and the format of historic volatility information affect risk perception and investment decisions. Risk Analysis 25(3):597-609.

Whitestone Research and Jacobs Facilities Engineering. 2001. Implementation of the Department of Defense Sustainment Model: Final Report. Washington, D.C. January.

Williams, B., and P. Nayak. 1996. A model-based approach to reactive self-configuring systems. Proceedings of the National Conference on Artificial Intelligence 2:971-978.

Wisconsin. 2009. Building Information Modeling (BIM) Guidelines and Standards for Architects and Engineers, Division of State Facilities, Department of Administration. Available at ftp:// doaftp04.doa.state.wi.us/master_spec/DSF%20BIM%20Guidelines%20&%20Standards/ BIM%20Guidelines%20and%20Standards.pdf. Accessed June 16, 2011.

Appendixes

Appendix A

Biosketches of Committee Members

DAVID A. SKIVEN, *Chair,* was a facilities management consultant and frequent adviser to federal agencies including the U.S. Navy and the U.S. Air Force. He was also codirector of the Engineering Society of Detroit Institute, a nonprofit organization dedicated to improving Michigan's economy. Mr. Skiven retired as the executive director of the General Motors Corporation Worldwide Facilities Group in 2007. The Worldwide Facilities Group was responsible for providing facilities management, utilities, construction, and environmental services, allowing General Motors (GM) clients to focus on their core businesses; this resulted in structural cost savings and improved use of assets. In 42 years at GM, Mr. Skiven worked in various engineering and plant operations, including as manager of facilities and future programs—manufacturing engineering for the Saturn Corporation and as director of plant environment and the environmental energy staff, before being appointed executive director of the Worldwide Facilities Group in 1993. Mr. Skiven served as a member of the National Research Council Board on Infrastructure and the Constructed Environment, on the board of directors of BioReaction, Inc., and on the board of the Engineering Society of Detroit. He had a BS in mechanical engineering from General Motors Institute and an MS degree from Wayne State University. Mr. Skiven was also a registered professional engineer. He was a member of the National Research Council Committee on Advancing the Productivity and Competitiveness of the U.S. Construction Industry.

GET W. MOY, *Vice Chair,* is a vice president of AECOM—a global design, management, and technical services company—and program director for AECOM's Federal Emergency Management Agency Public Assistance Technical Assistance Contract. Before joining AECOM, Dr. Moy served as an engineer for various sectors of the federal government, including the Naval Facilities Engineering Command and the Department of Defense (DOD). As director of utilities and energy in the Office of the Deputy Under Secretary of Defense (Installations and Environment), he was responsible for DOD's energy program, where he offered insight on such issues as the security of utility infrastructure, the role of distributed generation and renewable energy, energy and water-resource management, utility acquisition, and utilities privatization. As the director of installations requirements and management at DOD, he was responsible for the administration and direction of installations worldwide. Dr. Moy has managed complex programs for the federal government, including projects with stringent energy and environmental mandates. He was the recipient of the U.S. 2007 Presidential Rank Award for Meritorious Service. He received the National Institute of Building Sciences President's Award, which is presented to persons who have substantially improved the building process through government service. He is a fellow of the American Society of Civil Engineers, and a member of the United States Naval Institute, the Society of American Military Engineers, and the Tau Beta Pi Engineering Honor Society. Dr. Moy is a graduate of the Naval War College. He received a BS in civil engineering from the Catholic University of America and a master's degree and doctorate of science degree in engineering administration from the George Washington University.

MICHAEL A. AIMONE is vice president for strategy development for Battelle Memorial Institute's National Security Global Business. In that capacity, he leads energy and infrastructure strategy and market planning for Battelle's world-class science and technology support of the U.S. military services, defense agencies, and other federal clients engaged in the vital mission of national security. Mr. Aimone is a former member of the Senior Executive Service and retired from the U.S. Air Force after 39 years of combined military and civilian service. While with the Air Force, he was responsible to the chief of staff for leadership, management, and integration of Air Force civil engineering, logistics readiness, supply, transportation, and aircraft and missile maintenance.

BURCU AKINCI is a professor of civil and environmental engineering at Carnegie Mellon University. Her research is focused primarily on information technologies and the development of formalized model-based approaches for analysis of construction projects. Her work involves developing models to capture building-related data over time to support decision-making during construction planning and execution and facility management. She collects data by using emerging sensing and data-capture technologies. Dr. Akinci is a coauthor of numerous articles in peer-reviewed journals, including "Technological assessment and process implications of field data capture technologies for construction and facility/infrastructure management," "Tracking components and maintenance history within a facility utilizing radio frequency identification technology," and

"Capturing and representing construction project histories for estimating and defect detection." She holds a BS in civil engineering from Middle East Technical University in Turkey and an MS and a PhD in civil and environmental engineering from Stanford University.

ALFREDO H-S. ANG is a research professor and professor emeritus at the University of California, Irvine. Since 1988, he has also been a professor emeritus at the University of Illinois at Urbana-Champaign (UICC), where he received his PhD and was on the civil engineering faculty from 1959 through 1988. He received his BS in civil engineering from the Mapua Institute of Technology and an MS in structural engineering from UICC. Dr. Ang has published about 400 papers and articles and a two-volume textbook on probability concepts in engineering. He is active in several engineering societies particularly the American Society of Civil Engineers (ASCE), in which he served as an international director on the Board of Directors from 1998 to 2001. He is the ASCE representative to the Asian Civil Engineering Coordinating Council and a member of the International Activities Committee. He is also a fellow of the American Society of Mechanical Engineers, an associate fellow of the American Institute of Aeronautics and Astronautics, a founding member of the International Association for Structural Safety and Reliability, honorary president of the International Association for Life-Cycle Civil Engineering, and a member of several other professional and scientific societies. He was elected to the National Academy of Engineering in 1976 for developing practical and effective methods of risk and reliability approaches to the formulation of engineering safety and design structural criteria.

JOSEPH BIBEAU is the president of Eagle Enterprises of Tennessee, LLC, a company that provides consulting services for business organizational development, including real-estate property management and investment. Before joining Eagle Enterprises, Mr. Bibeau was the group director for energy and utility services for the Worldwide Facilities Group at General Motors (GM). In that position, he was responsible for utility procurement, engineering, conservation, powerhouse, and wastewater treatment operations for GM North America (GMNA), and he coordinated energy and utility activity for GM worldwide. He managed an annual budget of $850 million, 800 GM employees, and more than 200 contract engineers. During his tenure, GMNA reduced its water consumption by 46 percent and its energy consumption by 30 percent on a volume-adjusted basis; this amounted to an annual savings of $25 million for water and $215 million for energy. Earlier in his career, Mr. Bibeau was the superintendent of maintenance, facilities, and controls for Saturn Corporation and manufacturing director for a startup automobile assembly plant in Gujarat, India. He holds a BS degree in mechanical engineering from Kettering University and attended California State University's master's of business administration program.

IVAN DAMNJANOVIC is an assistant professor in the Zachry Department of Civil Engineering at Texas A&M University. His research focuses on construction-project development, finance and management and analytical models to support decision-making. His teaching interests are in construction-project management, contracting, operations-research methods, engineering economics, real options, and project finance. Dr. Damnjanovic is investigating construction-project complexities related to financial feasibility, energy conservation, environmental protection, and natural hazards mitigation. He holds a degree from the University of Nis, Serbia, and a PhD in civil engineering from the University of Texas at Austin.

LUCIA E. GARSYS is the deputy county administrator for development and infrastructure for Hillsborough County, Florida. She manages 1,800 employees and a $550 million, 6-year capital program. She is responsible for managing the life cycle of transportation, stormwater, water, and wastewater systems and more than 500 government facilities, including fire stations, libraries, parks, courts, and office buildings. Ms. Garsys directed initiatives to create a preservation and maintenance program for facilities. She is identifying alternative ways of delivering local government services in an effort to consolidate and eliminate facilities. Ms. Garsys has 30 years of public-sector and private-sector experience, including capital and asset management, planning, fiscal-impact analysis, development and redevelopment using tax-increment financing and organizational and process improvement. She is a member of the American Institute of Certified Planners. Ms. Garsys served on the National Research Council Board on Infrastructure and the Constructed Environment from 2004 to 2009 and on the Committee on Business Strategies for Public Capital Investment. She holds a bachelor's degree in city and regional planning from the Illinois Institute of Technology and a master's degree in urban planning from the University of Illinois at Urbana-Champaign.

DANIEL F. GELDERMANN is a principal analyst at Calibre Systems, Inc., a firm specializing in management and technology services. Mr. Geldermann has more than 27 years of experience that includes directing all aspects of facilities-engineering management—planning, engineering, design, contracts, operations, maintenance, repair, construction, utilities, environmental, transportation, safety, real estate, historic properties, and family housing—at various locations in the United States, Asia, and Europe. In addition to his consulting experience, his expertise has been developed through a career as a U.S. Navy Civil Engineer Corps officer and as an associate director of facilities at a state university. Mr. Geldermann has managed facility-related organizational budgets, service-contract programs, construction-management projects, facilities planning, commissioning services, and facilities operations. As a consultant, he has conducted numerous facility-management studies and reviews for agencies, including the Smithsonian Institution, the Department of Health and Human Services, the U.S.

Army, and the National Aeronautics and Space Administration. Mr. Geldermann is a registered professional engineer in Wisconsin and Virginia and a certified facility manager, and he holds a master facility executive certificate from the Building Owners and Management Institute. He is a past chair of the Society of American Military Engineers National Facilities Asset Management Committee. Mr. Geldermann holds a BS in civil engineering from Marquette University and an MS in financial management from the U.S. Naval Postgraduate School.

MICHAEL R. GREENBERG is a professor, associate dean of the faculty, and director of the National Center for Neighborhood and Brownfields Redevelopment, and director of the National Transportation Security Center of Excellence at Rutgers, the State University of New Jersey. Dr. Greenberg studies environmental health and neighborhood redevelopment policies. His books include *Urbanization and Cancer Mortality* (1983), *Hazardous Waste Sites: The Credibility Gap* (1984), *Public Health and the Environment* (1987), *Environmental Risk and the Press* (1987), *Environmentally Devastated Neighborhoods in the United States* (1996), *The Reporter's Environmental Handbook* (2003), and *Environmental Policy Analysis & Practice* (2008). He has been a member of National Research Council committees that focus on waste management, such as the destruction of the U.S. chemical-weapons stockpile and nuclear weapons. He has received awards for research from the Environmental Protection Agency, the Society of Professional Journalists, the Public Health Association, the Association of American Geographers, and the Society for Risk Analysis. He serves as associate editor for environmental health for the *American Journal of Public Health* and is editor-in-chief of *Risk Analysis: An International Journal.* Dr. Greenberg holds a BA from Hunter College and an MA and a PhD from Columbia University.

WILLIAM G. STAMPER is a consultant and chief executive officer of CBC Solutions, Inc., a facilities-management consulting firm. He retired from the federal government in 2007 as the deputy assistant secretary for facilities management and policy at the Department of Health and Human Services. In that capacity, he reestablished an office to lead departmental efforts related to real property, facility engineering, environmental management, historic-preservation, headquarters operations, security, and safety. During his government career, Mr. Stamper also worked at the National Aeronautics and Space Administration (NASA) in a variety of positions, including headquarters facility program manager for a variety of NASA centers, national aeronautics facility manager, and program manager for the $2.6 billion National Wind Tunnel Program; he finished his tenure as NASA's deputy director of facilities. Early in his career, Mr. Stamper worked at the Air National Guard (ANG), where he was responsible for planning, project development and approval, and submission of the ANG military construction budget to the Office of Management and Budget and Congress.

ERIC TEICHOLZ is president and founder of Graphic Systems, Inc., a Cambridge, Massachusetts, firm specializing in facility management and real-estate automation consulting, system integration, market research, education, and publishing. He is a fellow of the International Facility Management Association (IFMA's highest honor), a member of the IFMA Foundation's Board of Trustees, chair of IFMA's Sustainability Committee, coeditor of the *International Journal of Facility Management,* and a member of the Facility Maintenance and Operations Committee at the National Institute of Building Sciences. Mr. Teicholz has helped organizations to define and implement technology for more than 25 years. He lectures internationally and is the author of hundreds of articles on computer graphics, facility management, computer-aided design and architecture, computer-aided facilities management and geographic information system technology. He is also the author or editor of 11 books on those subjects. Mr. Teicholz was educated as an architect at Harvard University. Before founding Graphic Systems, he spent 12 years at Harvard's Graduate School of Design as an associate professor of architecture and associate director of Harvard's largest research and development facility, the Laboratory for Computer Graphics and Spatial Analysis.

DONALD R. UZARSKI has been on the University of Illinois at Urbana-Champaign (UIUC) civil-engineering faculty since 1994. He retired in 2004 from the U.S. Army Engineering Research and Development Center-Construction Engineering Research Laboratory (ERDC-CERL) after 20 years of service. At ERDC-CERL, Dr. Uzarski conducted research to develop the science of facilities asset management, including modeling the decision-making process, determining the data required to support decisions, establishing business rules to support the process, creating new metrics to measure condition and performance, and performing necessary analyses. He served as a principal investigator and project manager for research efforts in railroad-track and building-asset management. He also served as a technical consultant in those fields to the U.S. Army and the U.S. Navy. Before his ERDC-CERL career, he served in various public-works assignments as a U.S. Navy Civil Engineer Corps officer. Dr. Uzarski earned his BS, MS, and PhD in civil engineering from the University of Illinois. He is the author of more than 70 papers, reports, and articles on the various aspects of infrastructure (railroads, roads, and buildings) asset management. Dr. Uzarski is a member of the editorial advisory board for the American Society of Civil Engineers (ASCE) *Journal of Infrastructure Systems* and an active member of ASCE, the American Railway Engineering and Maintenance-of-Way Association (AREMA), and the National Research Council Transportation Research Board (TRB). He serves or has served on several national committees for ASCE, AREMA, and TRB and is a past chair of the ASCE Infrastructure Asset Management Committee. He is a registered professional engineer in Illinois and Pennsylvania.

Appendix B

Committee Interviews and Briefings

2009

December 14   Peter Marshall, Chair, Federal Facilities Council
William F. Broglie, National Oceanic and Atmospheric
Administration
Karl Calvo, U.S. Coast Guard
Joseph Morganti, U.S. Air Force
John Yates, Department of Energy, Office of Science
Peter O'Konski, Department of Energy, Office of Engineering
and Construction Management
Dino Herrera, Department of Energy, National Nuclear
Security Administration
Alex Willman, Department of State
Steven Beattie, Department of the Navy
Dominic Savini, Federal Accounting Standards Advisory Board

2010

February 18   Terrell Dorn, Government Accountability Office (GAO)
Maria Edelstein, GAO
Peter Lufkin, Whitestone Research
Jay Janke, Whitestone Research
Lander Medlin, Association of Higher Education Facilities
Officers-APPA
Douglas Christensen, Center for Facilities Research-APPA
James J. Dempsey, U.S. Coast Guard

Al Antelman, U.S. Navy
Lance Marrano, U.S. Army
Michael Grussing, U.S. Army

February 19   William McNab, General Motors
Robert St. Thomas, IBM Corporation
Patrick Okamura, General Dynamics
Carl Rabenaldt, Parsons Corporation

May 11   Gerald Kokos, VFA, Inc.
Stephen Wooldridge, U.S. Army Health Facility Planning
    Agency
Cynthia Vallina, Office of Management and Budget
Douglas Ellsworth, U.S. Army Corps of Engineers
Andrew Dichter, U.S. Air Force (retired)
Valerie Baldwin, Baldwin Consulting
Kim Toufectis, National Aeronautics and Space Administration

Appendix C

Some Fundamentals of the Risk-Based Approach

BASIC PRINCIPLES

The fundamental tools needed for the quantitative risk-based approach to decision-making include the basic principles of probability. Those principles start with the premise that in the presence of uncertainty, a phenomenon or physical process can be defined or represented by a *random variable* and its *probability distribution*. That is, uncertainty is modeled as a random variable with a range of possible values and their probabilities defined by a probability distribution.

Thus, if X is a random variable with a range of possible values from a to b, its probability distribution may be defined as $F_x(x) = P(X \leq x)$; $a \leq x \leq b$.

Within the range of possible values of a random variable, there will be a *mean* (or average) value and a measure of dispersion, such as the *variance* or *standard deviation*. The ratio of the standard deviation to the mean is the *coefficient of variation* (COV).

Among the useful probability distributions are the *normal* or *Gaussian* distribution and the *lognormal* (or logarithmic normal) distribution.

The Normal or Gaussian Distribution. The normal distribution, whose range of possible values is $-\infty$ to $+\infty$ is denoted as $N(\mu, \sigma)$ where μ is its mean value and σ is its standard deviation. If $\mu = 0$ and $\sigma = 1.0$, the distribution is called the *standard normal distribution*. For the standard normal distribution, the probability from $-\infty$ to x is $F_x(x) = \Phi(x)$, where $\Phi(x)$ is tabulated in *Tables of Standard Normal*

Probability. The probability of a random variable, X, between a and b can be evaluated as

$$P(a < X \leq b) = \Phi\left(\frac{b - \mu_x}{\sigma_x}\right) - \Phi\left(\frac{a - \mu_x}{\sigma_x}\right),$$

where μ_X and σ_X are, respectively, the mean and standard deviation of X.

The Lognormal Distribution. In the lognormal distribution, whose range of possible values is 0 to ∞, there are no negative values. The probability that X will be between a and b becomes

$$P(a < X \leq b) = \Phi\left(\frac{\ln b - \lambda_x}{\zeta_x}\right) - \Phi\left(\frac{\ln a - \lambda_x}{\zeta_x}\right),$$

where λ_X and ζ_X are, respectively, the mean and standard deviation of $\ln X$ and they are the parameters of the lognormal distribution. These parameters are related to the mean and standard deviation of X as follows:

$$\lambda = \ln \mu_X - \frac{1}{2}\zeta^2$$

and

$$\zeta^2 = \ln\left[1 + \left(\frac{\sigma_X}{\mu_X}\right)^2\right] = \ln\left(1 + \delta_X^2\right)$$

if the COV of X, δ_X, is not large, say < 40%, $\zeta \cong \delta._X$.

MATHEMATICS OF PROBABILITY

A few rules that pertain to the mathematics of probability may be described briefly as follows.

Probability is defined with reference to the occurrence (or nonoccurrence) of an *event*, and for an event E

$$0 \leq P(E) \leq 1.0.$$

The Addition Rule. For two or more events, A and B, the "union" of A and B, denoted $A \cup B$, means the occurrence of A or B (or both), and the probability is given as the *addition rule*. namely,

$$P(A \cup B) = P(A) + P(B) - P(AB),$$

in which $P(AB)$ stands for the simultaneous occurrence of A and B.

The Multiplication Rule. The probability of the simultaneous occurrence of two events, A and B, is given by the multiplication rule, namely,

$$P(AB) = P(A|B){\cdot}P(B); \text{ or}$$
$$= P(B|A){\cdot}P(A)$$

in which $P(A|B)$ stands for the "conditional probability" of A given (or assuming) the occurrence of B.

Those two simple rules, together with the "theorem of total probability" and the "theorem of Bayes," constitute the basic rules of the mathematics of probability. For a more complete description of the theory of probability and illustrations of its many applications in engineering, see Ang and Tang (2007).

ILLUSTRATIVE APPLICATIONS TO SPECIFIC OUTCOMES

Described below are the numerical calculations of the risk or probability of "negative benefits" of three specific outcomes—accident rates and types, deferred maintenance, and energy use.

Accident Rates and Types

In this example, let

X = recordable incident rate (RIR),
Y = lost-time incident rate (LTR), and
Z = number of worker compensation claims.

Assume that the current incident rates and claims are as follows:

X = 4 per 100,000 hours
Y = 0.5 per 100,000 hours
Z = 1.

The average cost per incident is $75,000, whereas that of a worker compensation claim is $100,000.

With an investment of $200,000 for maintenance and repair, the incident rates would be reduced as follows:

X' = 2 per 100,000 hours
Y' = 0.1 per 100,000 hours
Z' = 0.2.

In this case, the current cost of an incident is

$$C = c_1 X + c_2 Y + c_3 Z,$$

and the corresponding reduced cost is

$$C' = c_1 X' + c_2 Y' + c_3 Z',$$

where c_1, c_2, and c_3 are the corresponding costs in dollars.
The pertinent costs are, therefore, as follows:

Current cost, $C = 4 \times 75,000 + 0.5 \times 75,000 + 1 \times 100,000 = \$437,500$.
Reduced cost, $C' = 200,000 + 2 \times 75,000 + 0.1 \times 75,000 + 0.2 \times 100,000$
$= \$377,500$.

The benefit derived from the investment in maintenance and repair, therefore, would be

$$\text{Benefit} = C - C'.$$

In this case, the benefit of maintenance and repair investment = \$437,500 − \$377,500 = \$60,000.

In this example, the risk that the investment will be greater than the savings (negative benefit) is $C < C'$. Because there are uncertainties in all the variables X, X', Y, Y', Z, and Z', there is some probability of negative benefit. For example, suppose that the uncertainties are ±30% in all the variables. The risk would be calculated as follows.

Assume that the variables are independent normal random variables; the means and standard deviations of each of the variables are

$$X = N(4, 1.2); Y = N(0.5, 0.15); Z = N(1, 0.3);$$

and $X' = N(2, 0.6)$; $Y' = N(0.1, 0.03)$; $Z' = N(0.2, 0.06)$.

The respective means of C and C' (assuming no uncertainties in the costs) are

$$\mu_C = \$437,500 \text{ and } \mu_{C'} = \$377,500,$$

whereas the standard deviations are

$$\sigma_C = 75000\sqrt{1.2^2 + 0.15^2 + 0.3^2} = 1.246 \times 75000 = \$93,450$$

and

$$\sigma_{C'} = 75000\sqrt{0.6^2 + 0.03^2 + 0.06^2} = 75000 \times 0.604 = \$45,300.$$

Therefore, the risk of negative benefit would be

$$P(C < C') = P[(C < C') < 0] = \Phi\left(\frac{0 - (437,500 - 377,500)}{\sqrt{93,450^2 + 45,300^2}}\right)$$

$$= \Phi\left(\frac{-60,000}{103,850}\right) = \Phi(-0.58) = 1 - \Phi(-0.58) = 1 - 0.719 = 0.281.$$

That means that with the investment of $200,000 in maintenance and repair, the risk of negative benefit will be about 28 percent.

Deferred Maintenance

Any equipment or facility has a finite and variable operational life. In realistic terms, the operational life may be represented as a random variable and described with a probability distribution. The probability distribution often used for this purpose is the *lognormal* distribution.

The Risk Problem

Consider the maintenance problem of air-conditioning (A/C) units. Assume that the operational life T of a typical A/C unit can be described with the lognormal distribution, with a median life of t_m months or years, and a COV of δ_T (or a standard deviation of $\sigma_T \approx \delta_T \times t_m$).

Suppose further that the current maintenance schedule calls for inspection and repair (if necessary) of an A/C unit every n months or years. However, if inspection or repair is deferred beyond the schedule, what will be the reliability (probability of performance) of the A/C unit until the next scheduled inspection? And what would be the cost implication of deferring maintenance?

Solutions

Assume that the A/C unit has an operational life of t_m = 5 years, and a COV of δ_T = 0.30. The probability that the A/C unit will fail to perform within a life of t years is given by $P(T < t)$. With the lognormal distribution of the operational life T, the probability is

$$P(T < t) = \Phi\left(\frac{\ln t_m - \lambda}{\zeta}\right)$$

in which λ and ζ are the parameters of the lognormal distribution. The reliability is then $(1 - P)$.

Problem I. The probability that the operational life of an A/C unit will be less than 2 years is determined as follows. The parameters of the lognormal distribution λ and ζ are

$$\lambda \approx \ln t_m = \ln 5 = 1.61; \text{ and } \zeta \cong \delta_T = 0.30.$$

The required probability of failure (non-performance) in 2 years is

$$P(T < 2) = \Phi\left(\frac{\ln 2 - 1.61}{0.30}\right) = \Phi\left(\frac{0.69 - 1.61}{0.30}\right) = \Phi(-3.07)$$

$$= 1 - \Phi(3.07) = 1 - 0.9989 = 0.0011.$$

Therefore, the probability that a typical A/C unit will fail within a 2-year period is 0.11 percent. Its reliability of performance, therefore, is $(1 - 0.0011) = 0.9989 = 99.89$ percent.

Problem II. Suppose that the A/C units of an agency are scheduled for routine maintenance at 2-year intervals; this maintenance schedule should ensure a high performance reliability (of 99.89 percent) However, because of circumstances (such as a shortage of technicians, or a shortage of funding), the inspection and repair schedule is deferred for 2 years (until the next scheduled maintenance).

The average cost of repair per A/C unit is estimated to be $1,500; the cost implication of the deferred maintenance will be as follows.

In this problem, the operational life is assumed to be longer than 2 years (the schedule for maintenance), so it is necessary to calculate the probability of failure in 4 years (2 years beyond the scheduled maintenance). The solution requires conditional probability as outlined below:

$$P(T \leq 4 | T > 2) = \frac{P[(T \leq 4) \cap (T > 2)]}{P(T > 2)} = \frac{P(2 < T \leq 4)}{P(T > 2)}$$

where, from Problem I,

$$P(T > 2) = 1 - 0.0011 = 0.9989,$$

$$P(2 < T \leq 4) = \Phi\left(\frac{\ln 4 - 2.30}{0.30}\right) - \Phi\left(\frac{\ln 2 - 1.61}{0.30}\right)$$

$$= \Phi\left(\frac{1.39 - 1.61}{0.30}\right) - \Phi\left(\frac{0.69 - 1.61}{0.30}\right) = \Phi(-0.73) - \Phi(-3.07)$$

$$= 1 - \Phi(0.73) - (1 - \Phi(3.07)) = \Phi(3.07) - \Phi(0.73) = 0.9989 - 0.7673 = 0.23,$$

and

$$P(T \leq 4 | T > 2) = \frac{0.23}{0.9989} = 0.23.$$

Therefore, deferring the maintenance of the A/C units for 2 years, or until the next scheduled maintenance, will result in a probability of failure of a typical A/C unit of 23 percent.

If the agency has 1,000 A/C units, 230 of them are likely to fail within 2 years beyond the scheduled maintenance. If the average repair cost is $1,500 per unit, the deferred maintenance cost will be $230 \times 1500 = \$345,000$.

Energy Use

Problem

Determine the benefit of investments in maintenance and repair of energy systems. Consider savings in oil equivalents (gallons) of gasoline consumption at a price of $3.00 per gallon. Assume that with an investment of I (dollars) the reduction in gasoline consumption is $Y = f(I)$; this function may have to be developed empirically from historical data.

Let the current consumption be X gallons; and in dollars = $3X$.

Assume that with an investment of I dollars, the reduced consumption would be Y gallons, and in dollars = $3Y$.

Therefore, the energy saving with investment I is $(X - Y)$ gal; or in dollars is $(3X - 3Y)$.

Hence, failure in this case may be defined as "saving (in dollars) is less than the investment"; that is, in dollars,

$$[3(X - Y) - I] < 0.$$

There will be uncertainty in X (the current consumption) and in Y (the reduced consumption), so there will be a probability of failure, or risk that the investment will be greater than the savings (negative benefit). To calculate that probability, assume that X and Y are both normal (or Gaussian) random variables, with respective means and standard deviations μ_X, μ_Y, and σ_X, σ_Y; i.e., denoted as

$$N(\mu_X, \sigma_X) \text{ and } N(\mu_Y, \sigma_Y).$$

The probability of failure, P, is

$$P = P[3(X - Y) - I] < 0 \text{ or } P[3(X - Y) < I].$$

For normal random variables, it becomes

$$P = \Phi\left(\frac{3(\mu_X - \mu_Y) - I}{\sqrt{(3\sigma_X)^2 - (3\sigma_Y)^2}} \right).$$

For numerical illustrations, assume hypothetically the following. Current average gasoline consumption is $\mu_X = 10$ million gallons, with a standard deviation of $\sigma_X = 2$ million gallons. With an investment of $10 million, the average reduced consumption is expected to be $\mu_Y = 8$ million gallons, and $\sigma_Y = 2$ million gallons. The risk of a negative benefit is

$$\Phi\left[\frac{3(10-8)-10}{\sqrt{6^2 + 6^2}} \right] = \Phi\left(\frac{-4}{\sqrt{72}} \right) = \Phi(-0.47)$$

$$= 1 - \Phi(0.53) = 1 - 0.70 = 0.30.$$

Therefore, with an investment of $10 million, there is a 30 percent probability of a negative benefit.

Determining the investment needed to reduce the risk to 10 percent is calculated as follows:

$$\Phi\left[\frac{3(10-8)-I}{\sqrt{6^2 + 6^2}} \right] = 0.10$$

$$\frac{6-I}{\sqrt{72}} = \Phi^{-1}(0.10) = -\Phi^{-1}(0.90) = -1.28.$$

The investment needed will be I = $16.86 million.

A Related Problem

Suppose that an agency wishes to invest $20 million to reduce expected gasoline consumption to $\mu_Y = 5$ million gallons, and $\sigma_Y = 1.5$ million gallons. The risk would be calculated as

$$\Phi\left[\frac{3(10-5)-20}{\sqrt{6^2 + 4.5^2}} \right] = \Phi\left(\frac{-5}{\sqrt{56.25}} \right) = \Phi(-0.67)$$

$$= 1 - \Phi(0.33) = 1 - 0.63 = 0.37.$$

In this case, the probability of a negative benefit is 37 percent, which may be too high.

To reduce the level of risk of a negative benefit to 10 percent, the investment needed would be calculated as

$$\Phi\left[\frac{3(10-5)-I}{\sqrt{6^2+4.5^2}}\right]=0.10$$

$$\frac{15-I}{\sqrt{56.25}}=\Phi^{-1}(0.10)=-\Phi^{-1}(0.90)=-1.28$$

$$\frac{15-I}{7.5}=-1.28;\ \text{or}\ I=9.6+15=24.6.$$

In this example, $24.6 million would be required to reduce the expected gasoline consumption to 5 million gallons with a 10 percent probability that the investment will be greater than the savings.

RELIABILITY ANALYSIS

The practical approach to ensure the reliability or safety of an engineered system is to apply the *first-order reliability method* (FORM). The basics of the method can be described below.

The evaluation of reliability of an engineered system may be considered as a problem of *supply* versus *demand;* for this purpose, define the following random variables:

X = the supply and
Y = the demand.

The objective of a reliability analysis is to ensure that $(X > Y)$.

If the probability density functions (PDFs) of X and Y are, respectively, $f_X(x)$ and $f_Y(y)$, then the reliability of the system is measured by the probability of failure (nonperformance),

$$p_F = P(X \le Y) = \int_0^\infty F_X(y)f_Y(y)dy,$$

in which

$$F_X(y) = \int_0^y f_X(x)dx.$$

The above is a convolution integral, shown graphically in Figure C.1.

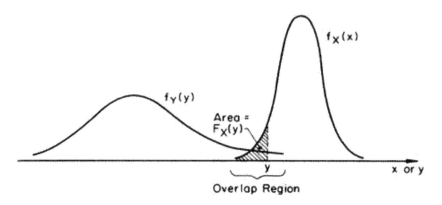

FIGURE C.1 The probability of failure.

The corresponding probability of performance is then

$$P_S = 1 - P_F.$$

Consider a system in which the available supply, X, is a Gaussian or normal random variable $N(\mu_X, \sigma_X)$ and the demand is also a Gaussian random variable $N(\mu_Y, \sigma_Y)$. The difference, $M = X - Y$, called the *safety margin*, is also a Gaussian variable with a mean of

$$\mu_M = \mu_X - \mu_Y.$$

If X and Y are statistically independent, the variance of M is

$$\sigma_M^2 = \sigma_X^2 + \sigma_Y^2.$$

Furthermore, $\dfrac{M - \mu_M}{\sigma_M}$ is $N(0,1)$. Hence, the probability of nonperformance is

$$P_F = F_M(0) = \Phi\left(\frac{-\mu_M}{\sigma_M}\right) = 1 - \Phi\left(\frac{\mu_M}{\sigma_M}\right)$$

in which Φ is the cumulative probability of the standard Gaussian distribution, $N(0,1)$.

Clearly, the reliability of the system is a function of the ratio $\mu M/\sigma M$, which may be called the *safety index* or *reliability index* and denoted by β. In this case,

$$\beta = \frac{\mu_M}{\sigma_M} = \frac{\mu_X - \mu_Y}{\sqrt{\sigma_X^2 + \sigma_Y^2}}.$$

If the supply and demand are both lognormal random variables, the corresponding reliability index would be

$$\beta = \frac{\ln(x_m/y_m)}{\sqrt{\delta_X^2 + \delta_Y^2}}$$

and the probability of nonperformance can be expressed as

$$p_F = \Phi(-\beta) = 1 - \Phi(\beta).$$

In the first case, where X and Y are both Gaussian random variables, the quantitative relation between p_F and β is unique (one to one), as follows:

p_F	β	p_F	β
0.5	0	0.01	2.33
0.25	0.67	10^{-3}	3.10
0.16	1.00	10^{-4}	3.72
0.10	1.28	10^{-5}	4.25
0.05	1.65	10^{-6}	4.75

The First-Order Reliability Method (FORM)

Engineers are traditionally reluctant to admit a probability of failure; for this reason, a good alternative strategy is to use an equivalent measure, the safety index β, which is a complete measure of the safety or performance of an engineered system.

This has served to spur the practical implementation of the probabilistic approach in engineering.

Using the β and FORM has contributed greatly to the practical implementation of reliability engineering (Ang and Cornell, 1974). The basics of FORM may be described as follows.

Introduce the reduced variates, X' and Y',

$$X' = \frac{X - \mu_X}{\sigma_X} \quad \text{and} \quad Y' = \frac{Y - \mu_Y}{\sigma_Y}.$$

In the space of X' and Y', the safe and failure states of the system may be represented as shown in Figure C.2.

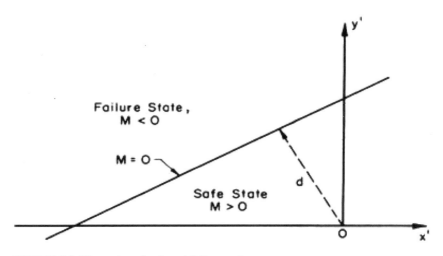

FIGURE C.2 Illustration of safe and failure regions.

In terms of the reduced variates, the limit state equation $M = 0$ ($X - Y = 0$) becomes

$$\sigma_X X' - \sigma_Y Y' + \mu_X - \mu_Y = 0.$$

From the above figure in the reduced variates, we can clearly distinguish the failure region from the safe region, and distinguish the limit state equation (or failure surface) that separates the two regions. On that basis, the distance, d, from the failure surface to the origin, o, is a measure of safety or reliability and in fact is the safety index β. That distance is (from analytic geometry)

$$\beta = \frac{\mu_X - \mu_Y}{\sqrt{\sigma_X^2 + \sigma_Y^2}},$$

and thus the probability of failure is

$$p_F = \Phi(-\beta).$$